THE SUN:
OUR FUTURE
ENERGY
SOURCE

THE SUN:
OUR FUTURE ENERGY SOURCE

DAVID K. McDANIELS
University of Oregon

JOHN WILEY & SONS
New York   Chichester   Brisbane   Toronto

Library of Congress Cataloging in Publication Data:

McDaniels, David K   1929–
  The sun, our future energy source.

  Includes bibliographies and index.
  1. Solar energy. I. Title.
TJ810.M17   621.47   78-12569
ISBN 0-471-04500-4

Printed in the United States of America

10 9 8 7 6 5 4 3 2 1

PREFACE

(Solar energy is certainly not a new concept. It has long been realized that, despite the low energy density of the incoming insolation, solar radiation has a large potential as an energy source. Although the possibility of electric power generation from solar radiation is very promising for the future, it is the application of solar energy techniques to the heating and cooling of buildings that lies immediately within the scope of present day state-of-the-art technology. The solar energy falling upon the roof of a typical residence is roughly 10 times as great as the annual heat demand for that house. Also almost 25 percent of our nation's energy consumption goes into space heating, space cooling, and water heating, which means that the rapid development of the solar heating (and cooling) industry could make a significant impact upon this country's energy budget by the turn of the century.)

(There is a substantial base of research upon which the application of solar energy to space heating and cooling can be built. Between 1940 and 1970 more than 25 houses and laboratory installations were partially heated with solar energy. The reason it has taken so long to implement the widespread use of solar energy is that while the solar fuel is free, a considerable capital investment is needed to harness it. The availability of cheap oil, electricity, and natural gas has seriously impeded solar energy applications in the past.) Only recently, because of the escalation in fossil fuel costs, has solar energy become economically competitive for space heating. The economic position of solar energy for space and water heating and for air conditioning should continue to improve in the next few years as the cost of conventional energy rises.

This book originated from lecture notes for a one-quarter introductory course on solar energy taught by the author at the University of Oregon. The course is one of several "minicourses" offered by the Physics Department to provide a liberal arts student with a broad background in scientific ideas and methods while dealing with topics that are of real interest to a wide spectrum of nonscience majors. Our success with these courses has more than justified the work required to prepare them. The enrollment in the course on solar energy increased from 25 to 450 students over a period of six years, reflecting the rapid increase of interest in this topic.

The principal aim of this text is to help the reader gain an understanding of the ways in which useful energy can be obtained from the incoming solar radiation. Thus, the major portion of this textbook covers topics such as solar radiation, flat-plate collectors, active and passive solar heating systems, solar thermal electric generating schemes, and photovoltaic devices. In order to round out the course and provide the requisite background, the first three chapters discuss topics related to the overall energy problem, such as the energy crisis, fossil fuel reserves, and nuclear fission. Some of this introductory material can be readily left out of the course, if desired, without seriously affecting the student's understanding of future developments.

In keeping with the introductory nature of the course, the presentation has been kept as nonmathematical as possible. The emphasis is entirely upon developing a qualitative understanding of the subject. Numerical examples are occasionally worked out to illustrate the principles involved, however students are not expected to be able to work out detailed calculations themselves. The problems at the end of each chapter were chosen to span a large range of difficulty. Some of the multiple-choice questions are exceedingly easy, included only to check that the reader is aware of the most basic facts. I hope the others will cause the interested reader to seriously think about the subject matter under discussion.

The system of units used in this book is the MKS system, which is essentially the SI system of units, with certain modifications. In this metric system the basic units of length, mass, and time are the meter, kilogram, and second. The corresponding metric unit of energy is the joule. Occasionally, English units have been used when it was felt that current practice demanded their usage.

Most people seriously interested in solar energy developments are also deeply concerned about energy conservation, recycling of valuable mineral resources, and pollution of our environment. Because of this concern, the book begins with a chapter concerning the question of energy growth and resource usage in a finite world. Chapter II introduces the student to the concepts and uses of energy. The resource potential of fossil fuels, geothermal energy, and nuclear fission is assessed in Chapter III. These chapters are intended to provide the necessary background about the present energy crisis, about the concept of energy, and about our ability to meet this crisis with our nonrenewable energy sources.

The coverage of solar energy begins in Chapter IV with a short, historical review and an assessment of the economic viability of solar energy for space heating as compared with our present energy alternatives. This discussion is followed by a slight diversion into cosmology in which the nature of the universe and the formation of the sun are described. Purists may object to the inclusion of this material, but our experience indicates a large student interest in how the sun was formed, how it compares with other stars, and how it produces the energy that we receive each day.

A rather extensive coverage of solar radiation is presented in Chapter VI. The concepts of wave motion, and light, the breakdown of the incident solar energy into direct and scattered components, and radiation monitoring instruments are among the topics examined. Also included is a brief discussion of blackbody radiation, the connection between emissivity and absorptivity, and Rayleigh scattering.

The last five chapters discuss the primary solar energy applications: space heating and cooling and solar electric power generation. Chapter VII begins with a simplified description of how flat-plate collectors work. The basic concepts relating to efficient collection design, heat loss analysis, details of construction, and selective surfaces are discussed. The following chapter integrates the flat-plate collector into the overall heating system. A solar heating system requires a number of unique features by contrast with conventional systems. Topics covered include heat-exchange fluids, heat storage, delivery and control of the collected solar heat, solar cooling, and the use of reflector enhancement. A discussion of active and passive solar houses is contained in Chapter IX. While active solar systems have been emphasized in this book in order to better explain the physical principles involved, it is quite likely that passive approaches will prove to be the most economic way to utilize solar energy for space heating.

While technically feasible right now, the economic development of solar electric power plants will require a massive and sustained program of research. The most promising ways to do this using thermally driven generating devices are reviewed in Chapter X. All of these approaches involve the use of a heat engine of some type. Hence, the chapter begins with an elementary discussion of heat and thermodynamics and the restrictions upon operating efficiency that result because of the second law of thermodynamics. The solar electric approaches covered include the use of a large field of tracking mirrors to reflect the sunlight upon a central receiver, intermediate-temperature solar energy collection using a large array of linear focusing collectors, and the generation of electricity using ocean temperature differences.

Another promising way to convert solar energy to electricity is with the aid of solar cells. These devices have been used in the space effort since 1958 and, in their early stage of development, the cost was of the order of $10 million per peak kilowatt. Improvements in manufacturing techniques have reduced this to less than $15,000 per peak kilowatt in 1977, which illustrates the considerable progress made in this area. However, at this price, solar cells are still prohibitively expensive for large-scale use. Basic principles of photovoltaic devices as well as promising lines of research and development are discussed in Chapter XI.

It is impossible to acknowledge fully the many sources of information that I have utilized in writing this book. A short, annotated bibliography of some of the more important and useful references are listed at the end of each chapter. Special thanks are owed to Dr. W. A. Shurcliff for contributing numerous, useful suggestions. I gratefully acknowledge many useful discussions and assistance from my colleagues at Oregon: Professors D. H. Lowndes and John Reynolds, and Steve Baker and Dan Kaehn. A great deal of stimulating and rewarding feedback was provided by the questions and remarks from the students in my classes. Their input played an important role in shaping the coverage and level of this book. Finally, this textbook could not have been written without the extensive secretarial assistance provided by Sandy Hill and the extensive secretarial and editorial assistance of my wife, Pat.

David K. McDaniels

CON-
TENTS

APPENDIXES

Our Finite World

1

Abundant and relatively inexpensive energy has, in a sense, built the world in which we live. Our tremendous industrial development, the conversion of raw, natural resources into finished products, our abundance of material goods, our food supply and agricultural practices, and the conveniences of our everyday lives all depend upon a ready supply of energy. The extent to which all facets of our society have become dependent upon energy can be illustrated by looking at the changes in agricultural practices over the last 60 years. Figure 1-1 shows graphically the rapid development of mechanized farming since 1900. The horse is no longer used as a prime mover. In 1971 the average American consumed about the same amount of food as in the 1940s, but this food is now grown using less manpower on less land, through the intensive use of fertilizers and machinery powered by fossil fuels. Food is shipped in elaborate, disposable wrappers and reaches consumers in all parts of the country while still fresh, because it is increasingly shipped by truck rather than by rail. What all of these steps have in common is that they are energy intensive. As another example, consider the rapid increase in the use of aluminum cans. It takes about four times as much energy to manufacture the metal for an aluminum can as for a steel one. From

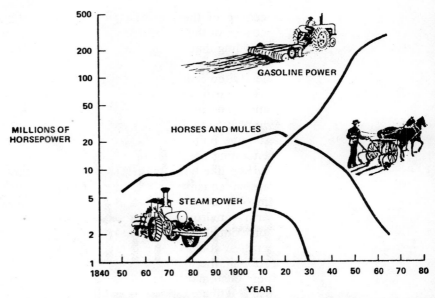

FIGURE 1-1   Depicts the growth of mechanized agriculture in this century. Oil has essentially replaced the horse as the prime mover on farms. (Courtesy of Harrison Brown. Reproduced, with permission, from the *Annual Review of Energy,* Vol. *1.* Copyright © 1976 by Annual Reviews Inc.)

an energy standpoint it would be more logical to use returnable glass containers, which are less energy intensive, than steel ones.

A.
GROWTH OF
ENERGY USAGE

The United States has enjoyed an abundance of raw energy resources including wood, coal, oil, hydropower, natural gas, and uranium. But our consumption of these energy resources has been reckless. With 6% of the world's population, the U.S. share of total world energy use is roughly 30%. [The U.S. energy usage was about 7×10^{19} joules in 1976.] Our per capita energy consumption is about four times that of Western Europe and 20 times that of Communist Asia. Over the course of the last five decades the use of energy in the United States in all forms has been doubling every 15 years. Until recently electrical energy production has been doubling every 9 to 10 years. This growth in energy usage has not varied simply in response to a corresponding growth of population. Energy usage in the United States, in particular, and in developed nations generally, has grown more rapidly than population, because of the simultaneously rising material standard of living and because of the continuous change to more energy-intensive methods of providing that standard of living. This is illustrated in Figure 1-2, which shows the per capita increase in energy and gross national product for the United States in the twentieth century.

The earth is a system with limited supplies of land, air, water, vegetation, and minerals. The rapid increase in energy use characteristic of the past 50 to 100 years cannot continue indefinitely in a finite system. At some point even the aesthetic damage to our lives from too many generating stations, too many transmission lines, too many factories, and the like becomes intolerable. One serious aspect of the increasing rate of consumption of our energy resources arises from the large amounts of waste heat given off. This waste heat could threaten the earth's environmental balance if our energy growth continues unheeded. For example, it has been estimated that a 2°F rise in the earth's temperature could melt the Arctic Ocean ice pack. As a more immediate problem, thermal pollution of some of our rivers has noticeably changed their customary ecological balance. Urban climatic conditions could also be affected. It has been estimated that if the present rate of increase of energy use in the Los Angeles area continues, a 5 to 7° rise in the ambient temperature will be experienced in the next 20 years.

The type of growth in energy resource consumption that we have witnessed throughout most of this century is exponential growth; growth characterized by the existence of a "doubling time," during which the

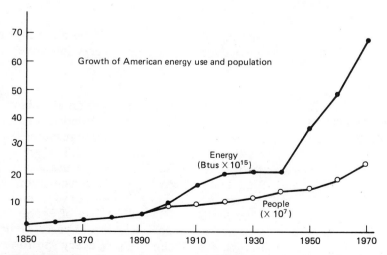

FIGURE 1-2 Since about 1940 the growth in consumption of our energy resource has increased more rapidly than the population.

energy consumed doubles, then redoubles again in another time interval equal to the doubling time, and so on. This exponential growth behavior is illustrated in Figure 1-3. As an example of exponential growth pushed to ridiculous extremes, consider that the amount of energy available from fusion of the deuterium in the oceans is sufficient (provided that controlled deuterium fusion reactors become available) to last for more than a billion years at our 1970 rate of consumption of energy. However, at our present doubling rate of 15 years, this practically infinite resource of energy would be consumed in 450 years. Of course, many other problems would arise long before this occurs.

Growth characterized by the existence of a doubling time eventually outstrips all of the finite resources in nature, and never continues unchecked in a finite system. The only naturally occurring instances of exponential increase are inevitably either transient (temporary) phenomena or are terminated by destruction of the system in which the growth takes place. Cancerous cell growth is one such naturally occurring example. The exponential increase of an animal population to a point where overgrazing and starvation of the entire population results, is

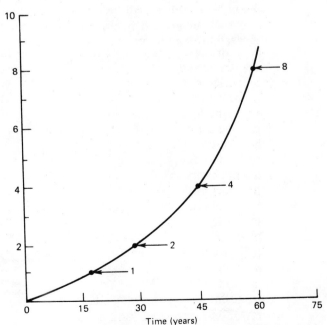

FIGURE 1-3   Graph illustrating the growth of energy usage with time assuming that the amount of energy consumed doubles every 15 years. This type of growth is called exponential growth.

another. The balance that can be reached between two populations of predator and prey animals shows how nature normally prevents exponential growth by limiting the increase in any one animal population to a size that can live in harmony with its surroundings. This type of balance, in physics, is called a "steady state" situation.

Only recently have doubts been raised about the legitimacy of our present and future energy needs. Estimates of future energy needs were (and still are) based largely on projections of existing rates of energy consumption; these projections are most often simply extrapolations into the future of present growth rates. Previously, no one questioned the right of economic growth and resultant exponential increase in energy consumption. Growth in terms of population, consumption of goods, improved technology, use of energy, and pollution has been an important factor in determining the character of our society. With the increasing concern for the environment and the growing realization that oil supplies are dwindling, the widsom of this pattern of behavior is now being questioned.

**B.
THE ENERGY
CRISIS**

Because our present energy supplies are both finite and nonrenewable our "energy splurge" appears to be coming to an end. Warning signs in the form of oil shortages and blackouts have been given. In the winter of 1971 Britain faced a serious lack of electric power, because of a nationwide strike of coal miners that lasted for a period of two months. Without electricity, activity in a modern nation soon lessens. The available electric power was rationed in some areas during blackouts for as little as nine hours a day. For several weeks the inhabitants of Britain were made starkly aware of the fact that electricity has become essential to modern life. More recently the domestic oil shortage brought home to the United States an awareness of the extent of our dependence upon depletable energy resources.

The heart of the present energy problem is that we have come to rely on oil for almost one-half of our energy. Table 1-1 shows a breakdown of the U.S. sources of energy for 1973. A detailed presentation of the U.S. resources is given in Chapter III, but the inescapable fact is that while one-half of our energy consumption comes from oil, our own production is now going down, with only a brief respite expected from the advent of Alaskan supplies. Generous estimates of how much oil might ultimately be extracted from the North Slope Fields indicate that they will only satisfy present U.S. petroleum needs for about a decade at most.

There are certainly a number of domestic energy resources that can be used to compensate for the impending shortages. Substantial quantities of oil have been left behind in "exhausted" reservoirs. This coun-

Table 1-1. Origin of the Energy Used in the United States During the Year 1973

Source	Percentage of Total
Coal	18.3%
Oil	47.7%
Natural gas	31.4%
Hydroelectric	1.3%
Nuclear	1.3%

try has great potential resources of oil shale. Our proved coal resources could last for several hundred years at the present consumption rate. If the breeder reactor proves feasible (and acceptable) our reserves of uranium could satisfy our needs for several thousand years. If it can be satisfactorily utilized, the United States has an enormous potential resource of geothermal energy. Finally, solar energy represents an almost inexhaustible source of energy.

Where then lies the problem? Harrison Brown, Professor of Economics at California Institute of Technology, aptly characterizes the situation by pointing out that we have become ensnared in an "energy trap," where inexpensive energy is taken for granted. According to Brown this has:

1. *Caused us to take extremely low energy costs for granted and to accept them as the norm to be expected for a very long time in the future.*

2. *Deterred the development of a broadly based industrial expertise in the use of alternative energy sources.*

3. *Caused us to give low priority to research programs aimed at developing alternative energy supplies.*

4. *Prevented us from investing adequate amounts of capital in known technological processes for utilizing alternative energy sources safely on a substantial scale.*

To this list we might add that this energy trap has hindered the development of adequate conservation measures.

Because energy is relatively cheap, it is used in many wasteful ways. The cost of the electricity used to produce common items such as paper, food, machinery, wood products, and transportation equipment is only 1 to 2 % of total production cost. It is no wonder that only recently have steps been taken to decrease the amount of energy used by industry, for heating and cooling residences, and for transportation. It is not our purpose here to analyze the reasons why cost for energy is as low as it is. But certainly federal subsidies and relaxed environmental regulations have contributed. Will higher energy costs in the future seriously change the basic economic structure of the country? Naturally, the answer to this question depends upon just how far the cost of energy escalates in the future. In Western Europe the price of energy is at least twice that of the United States. But, this region still maintains a modern industrialized society.

There are also factors affecting energy production that have more to do with political and environmental decisions than with dwindling natural resources. For example, the federal government may have accentuated the natural gas shortage through its pricing policies. Antipollution measures have curbed the use of high-sulfur coal, while the government has provided insufficient support and incentive to the research necessary for the clean utilization of our plentiful coal reserves. Construction of new nuclear power stations has been opposed on grounds such as safety and radioactive waste disposal. Offshore exploration for oil and gas has also been slowed by environmental concerns. These environmental concerns are important, but it must be recognized that satisfying environmental requirements will raise the price of a unit of energy.

Provided that public policy can cope with the political and environmental considerations, there would appear to be adequate energy resources for both the near- and long-term future. Probably very few of us wants to go back to the "good old days," if it means completely giving up the energy-consuming trappings of our daily lives. Central heating, TV, electric appliances, etc., have become part of a way of life. But the era of cheap, unlimited power is coming to a close. No longer will we be able to design and construct residential and commercial buildings, industrial projects, and transportation systems in the most convenient manner possible, without regard for how they fit into the overall energy use and conservation picture.

**C.
ENERGY
CONSERVATION**

What can be done to remedy our energy crisis? Clearly we need to continue development of energy resources other than petroleum. But this will take time. An immediate solution is to embark upon a large-scale program to reduce the amount of energy that is wasted. Until recently energy conservation* was virtually ignored or dismissed. This view is typified by a report of the Chase Manhattan Bank that maintains the thesis that any major reduction in energy usage would harm both the nation's economy and its standard of living. The point of view of this report is that most of the energy is used for essential purposes; two-thirds go into commercial business needs with most of the remainder providing for essential private needs. The major conclusion is that although some minor uses of energy could be regarded as strictly nonessential, their elimination would not permit any significant savings.

Most informed people do not share this viewpoint. For example, the American Institute of Architects has pointed out that by 1990, more energy could be saved by energy conservation techniques in buildings than could be obtained from any of the following sources: domestic oil, conversion of shale rock to oil, Alaskan North Slope oil, domestic and imported natural gas, or nuclear energy. Individuals waste energy and, as discussed in detail below, there is much each person can do to conserve. But it must be stressed that energy in this country has been wasted primarily because public and private policies have long promoted energy extravagance in the mistaken belief that this was the best way to achieve a prosperous economy.

The use of energy conservation techniques does not necessarily create a lower standard of living. The standard of living in Sweden is comparable to that in this country even though Swedes use less than one-half the energy per person used by Americans. According to Schipper (see bibliography at the end of the chapter) 24% of the energy used in thermal power plants in Sweden is utilized for low-temperature process or space heating purposes. In the United States this waste heat is almost entirely discarded to the environment. He also claims that an evaluation, taking into account the slightly smaller size of Swedish dwellings, the higher percentage of apartments, the higher efficiencies of the well-maintained apartment heating systems, and the Swedish climate shows that space heating requirements in Sweden are 30 to 50% lower per square foot of space in homes and commercial buildings than in the United States.

As a nation we have the potential to save almost a factor of 2 in our energy consumption through conservation without drastically curtailing our standard of living. Let us review briefly some of the possibilities. Major improvements can be made in the design and insulation of homes and commercial buildings. The most significant parameters for energy consumption by a residence are the ambient (outside) temperature and wind speed. Inadequate insulation and leakage of outside air into homes

---

* It must be emphasized that energy can neither be created nor destroyed. All that can be done is to transfer it from one form to another (for more detail, see Chapter II). In this chapter we use the expression, energy conservation, to mean reducing usage and waste of our energy resources.

increase the energy used for both heating and cooling. One study of model homes in three different regions of the country showed that additional insulation in walls and ceilings, proper use of weather stripping, and foil insulation in floors would save an average of 42% of the energy consumed for space heating. A similar study of commercial buildings estimates that savings up to 40% could be obtained in these structures as well. The potential savings from both could amount to more than 7% of our present national total energy use.

Most buildings are overheated. Savings of 15 to 20% can readily be obtained through the simple procedure of turning down thermostats. This is easily shown by the following example. A typical home in the Pacific Northwest might have an equilibrium temperature inside of 21 degrees Celsius (21°C) while the ambient temperature is 7°C. By lowering the inside average temperature to 18°C, which could easily be attained through operation at 21°C in the daytime and 15°C at night, a savings of

$$(1\text{-}1) \qquad\qquad 1 - \frac{18 - 7}{21 - 7} = 1 - 0.79 = 0.21$$

or 21% can be attained. Operation at these temperatures may even be better for one's health! More refined estimates of the heating season savings across the country give roughly this result.

A window that is not receiving sunlight is responsible for a tremendous loss of heat in cold weather. Even a tightly weather-stripped window of double glass loses 6 to 10 times as much heat as an equal area of well-insulated wall. But in sunlight, the same window is often a net gainer of heat. The sun's rays pass right through the glass and warm the floors and walls inside. Each sun-heated floor or wall then becomes in turn an emitter of infrared rays, but at a longer wavelength that will not pass back out through the glass. The solar heat is effectively trapped. A useful window design might appear as shown in Figure 1-4. This is an example of the use of passive solar heating, which is described more fully in Chapter IX.

Improvements in the efficiency of utilization of energy can be made through the better design and implementation of some of our common household devices. The difference in energy efficiency between various models of common air conditioners is more than a factor of 3. Even better, one can use low-energy fans for cooling. Furnaces for space heating are often operated at only about one-half their stated efficiency because of poor maintenance and design. Electrically driven heat pumps could possibly double heat energy output. Frost-free refrigerators and freezers use almost twice the energy of those having a manual defrosting system. Fluorescent lights use only a quarter as much electricity as incandescent bulbs (although the reradiated energy from either type does go into useful space heating).

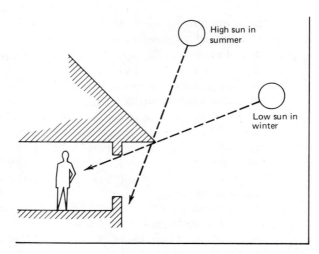

FIGURE 1-4   Proper overhang of eaves or awnings over window lets
winter sunshine in to provide solar heating, yet screens
out summer sun to help keep home cool.

Improved design of modern high-rise buildings could lead to savings of the order of 5% of the national energy budget. Many large buildings are so poorly designed that they require air conditioning to take away the heat generated by the excessive use of lights. The architect Richard Stein believes that careful design—using reflective window glass, orienting the building properly with respect to the sun, reducing air exchange and interior illumination—could reduce the operating energy by one-half over conventional design. The 1454-foot Sears and Roebuck building in Chicago requires more electricity than the city of Rockford, Illinois, with 147,000 people. A basic energy-wasting factor in building design is the use of large glass expanses as typified by the Lever House in New York City. Glass transmits heat to the outside readily in winter and admits solar radiation in the summer, overtaxing the building's air conditioning systems. The simplest solution is to substitute well-insulated walls for much of the glass area.

Overall savings of 10 to 20% could be made through improved industrial practices. The simplest way to achieve energy savings is to improve building insulation and reduce thermostats. Another way to obtain significant savings is to cascade high-temperature processes with lower-temperature demands. The whole area of utilizing waste heat for useful purposes is almost unexplored in this country. Significant savings can also be made through the use of improved industrial technology. The aluminum industry offers an example of this. Older processing plants require about 20,000 kilowatt-hours per ton (kW-hr/ton) while more efficient modern plants have cut this to about 9000 kW-hr/ton. It should also be noted that by recycling aluminum, the energy cost can be reduced to less than 2000 kW-hr/ton.

Savings in transportation may be harder to attain, since they will require very substantial changes in the American life-style. But the potential for energy conservation is considerable. When both direct and indirect energy costs are included, the auto accounts for almost 21% of the total U.S. energy consumption! A Stanford Research Institute report indicates that automobiles directly accounted for 13% of all U.S. fuel consumption in 1968, with trucks using another 5%, airplanes 2%, trains 1%, and buses 0.2%. Reducing the weight of automobiles would result in large savings as shown in Figure 1-5.

The movement of people and materials on the nation's highways accounts for 76% of all the energy consumed in transportation. Airplanes account for 10% of the remainder, with even smaller amounts for rail (3.5%), shipping (4.8%), and fuel pipelines (5.2%). In the last two decades the fastest growing method of moving people and goods has been by airplane, which is also the most energy intensive as shown in Tables 1-2 and 1-3. The airlines' share of total intercity passenger

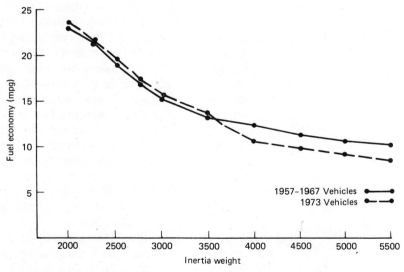

FIGURE 1-5   Disregarding minor efficiency fluctuations from model to model the fuel consumed per mile decreases remarkably as the car weight goes down. (Courtesy of Leo Schipper. Reproduced, with permission, from the *Annual Review of Energy,* Vol. 1. Copyright © 1976 by Annual Reviews Inc.)

Table 1-2. Energy Intensiveness for Passenger Transport in Units of Joules per Passenger Kilometer

Item	Urban	Intercity
Bicycle	0.339×10^6	
Walking	0.509×10^6	
Buses	6.28×10^6	2.71×10^6
Railroads		4.92×10^6
Automobiles	13.74×10^6	5.77×10^6
Airplanes		14.25×10^6

Source. E. Hirst, Oak Ridge National Laboratory.

traffic increased by a factor of 5 between 1950 and 1970, and their share of freight, sevenfold.

From an energy standpoint Table 1-2 shows that it makes good sense to move toward bus and railroad mass transit. What makes mass transit so efficient is its large load factor: the average number of passengers carried per mile of actual travel. With increased passenger traffic, the mass transit efficiencies in Table 1-2 could be further improved by at least a factor of 2. The decline of both rail service and urban mass transit relative to autos and airplanes reflects the greater convenience and the savings in time that are usually possible, at a high-energy cost, with these two modes of transportation.

Table 1-3. Energy Intensiveness for Freight Transport

Item	Joules/Ton-kilometer
Pipeline	0.763×10^6
Railroad	1.14×10^6
Waterway	1.15×10^6
Truck	6.44×10^6
Airplane	71.2×10^6

Source. E. Hirst, Oak Ridge National Laboratory.

D.
QUESTIONS OF
PHILOSOPHY

The behavior patterns of modern society have developed for a wide variety of political, historical, and economic reasons. There are a few common beliefs and philosophies that have strongly influenced these behavior patterns. For example, most of us are not yet truly aware of the long-term implications of the fact that the earth is a finite system. The most glaring example is that of oil. Figure 1-6 shows two estimates of the life cycle of the world's oil production. Q_∞ on this figure is the total available resource; the two quite different estimates of this quantity give a difference of only 10 years between the time of estimated peak production. This relationship between oil production and time was first pointed out over 20 years ago by the eminent geologist M. K. Hubbert. Little attention was paid at that time to his prediction that the U.S. production would peak in the 1970 to 1975 period. The actual oil production rate has completely vindicated this prediction.

Many of the common minerals may soon be in the same state of affairs. It is mainly in this century that the mineral resources have been tapped to provide the base for our phenomenal industrial and economic expansion. The metal consumed in 30 years at present growth rates

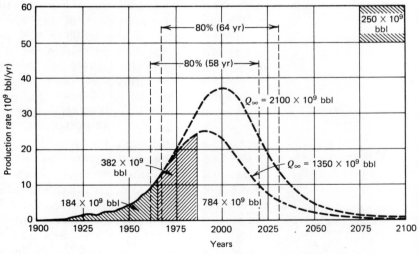

FIGURE 1-6   World's annual usage of oil plotted as a function of time. By the year 2000 the annual production of oil will have reached or passed the maximum. The "oil age" will only last a brief 100 years at most. The two different curves refer to different estimates of the total available world oil resource. [Originally published in the *Canadian Mining and Metallurgical Bulletin*, Vol. 66, No. 735, pp. 37–54 (1973).]

would be equal to that used in all previous history. Most of the world's prime mineral reserves have been found, and the United States and other countries have been mining them for years. By the end of this century the better grades of copper, tin, lead, mercury, and silver will have been seriously depleted. The mining of these better-grade ores has required little energy compared to the enormous drains that will be required to mine the low-grade, dispersed minerals in the future.

Shortages in other precious resources are imminent. Most everyone is now aware of the impending shortage of natural gas (see Chapter III), but few are aware that our helium reserves will soon run out. We presently use it in such wonderfully wasteful ways as providing lift for balloons. Our reserves of this precious commodity will be mostly gone by the turn of the century. Helium is a national asset for cooling objects to low temperatures at which some metals become superconductors of electricity. In the future, superconducting, electric-power transmission lines cooled by liquid helium may be highly desirable but it is questionable whether helium will be inexpensive enough to allow this technology to be implemented. Possibly, helium will be too expensive and too valuable to use outside of the research laboratory by the year 2000.

Little thought has been given to the needs of future generations. The world's supply of oil and natural gas was built up over a period of 600 million years by the conversion of solar energy through photochemical reactions. Society will use up 80 to 90% of this supply in 100 years leaving little left for the vital petrochemical industry. Our best fishing areas lie adjacent to the world's land masses. Continued pollution of these fishing areas with waste from the large population centers poses a real threat to the fishing industry. The excuse usually given for the rapid exploitation of our resources is that of practical economics. The time is near, however, when the nation's long-term needs and environmental considerations must be given as much weight in policy decisions as short-term, economic considerations.

The impressive accomplishments of modern science and technology have led most of us to believe that whenever a difficult problem such as the energy crisis arises, it will be overcome. It is certainly true that whenever the full power of our scientific and engineering talent is brought to bear upon a problem, the successes can be remarkable. The development of the atomic bomb during the Second World War and the Apollo flights to the moon are striking evidence of the ability of American science and technology to succeed with "crash programs" devoted to specific goals. On the other hand we tend to overlook the fact that the available supply of raw materials, talent, and resources was always far greater than was needed to accomplish these specific objectives. It is just as easy to overlook the vitally important, but slow, painstaking basic research that has gone into laying the foundation for these achievements. Consider the invention of the transistor, which came about only after a long series of developments, starting with basic solid-state theory in the 1930s and early practical solid-state devices in the late 1940s. Work on various types of associated devices continued in the early 1950s, particularly at the Bell Telephone Laboratory, finally culminating in the first transitor devices. The development came about as the natural consequence of a broad program of research, in many related areas and not as the result of a crash program designed to develop only a transistor. One cannot invent a transistor if its existence is not even known. Technological development must go hand in hand with a broad program of basic research.

An example of a problem which, thus far, has defied massive infusions of money and talent in the understandable hope for a quick solution, is cancer. Although much progress has been made, it has become quite apparent that this problem requires time-consuming fundamental research into the origins of cellular structure, the biochemistry of humans, the makeup of the body's immunological elements, and so on. Chances are that great advances will be made in the next 20 to 30 years, but this surely will not be easy.

Another extremely difficult problem is that of obtaining useful energy from fusion reactions. At one time this was looked upon eagerly as the solution to all future energy needs. After years of research it is evident that attainment of practical fusion power will probably not come for a long time. Work on fusion should probably be regarded as a great scientific adventure. Present research is still a long way from obtaining the requisite temperature and density of the plasma (a gas of charged particles) that is needed. Before practical fusion reactors can enter the engineering stage, basic research must reach the point where considerably more energy comes out of the fusion device than is pumped in. While those involved with fusion power generation are optimistic, and the reward of obtaining a practically infinite energy source provides great motivation, a long, difficult program of research lies ahead.

An even more difficult task is that of space travel to the stars. The nearest one is over 4 light-years away (1 light-year is the distance traveled by a beam of light, moving at 186,000 miles/sec, in one year, i.e., about 5.8×10^{12} miles) and is probably not sufficiently interesting to be worth visiting. The theory of relativity precludes travel at speeds equal to, or greater than, the speed of light. To get a rocket moving at speeds approaching that of light would require tremendous amounts of energy. The problems of maintaining living conditions during the trip, turning around to return, stopping upon returning, etc., seem impossible to solve at present. Furthermore, if we could somehow get near to the speed of light for most of the trip, the time span for the space travelers would be greatly reduced over that measured by someone on earth, so that when the space-trip travelers returned, every nontraveller would be much older. Present-day knowledge and technology is totally inadequate for this kind of a project.

Social considerations related to future energy sources are also important. While the potential harm to society of energy shortages is great, the increasing utilization of technologically complex energy sources will be associated with a variety of difficulties. Dangers associated with fossil fuel development include the risks of spills from pipelines and offshore oil development, reclamation of land that has been strip mined, and the environmental cost of shale oil development. Nuclear power implies a commitment to the management of waste products that long outlast the power plants themselves. These kinds of social issues must influence decision making about energy use, especially since they favor more efficient energy use than would be dictated solely from economics.

This book is primarily devoted to the subject of solar energy, our only important inexhaustible energy source. Its use offers the promise of an energy source that is both safe and environmentally sound. In its rush to plunder the earth of its dwindling reserves of nonrenewable resources, society has forgotten about our future generations. It is now time to plan for a society that operates in a framework of environmental and social stability. The day is past when the industrial world could think only of economic growth, with little concern for the needs of future inhabitants.

Several adverse features of solar energy must be considered in designing a particular application: (1) Its diffuse nature requires the use of relatively large surface areas for collection, (2) there are seasonal variations in the amount of energy available, (3) there is a daily variation of the available energy, which means that some means of storing energy must be available, and (4) there are short-term fluctuations in the incident insolation, which will make the available solar power somewhat irregular at times. The attractive features are that the sun will last forever (practically speaking), and its use entails a minimal degradation of the environment. The use of solar energy adds no heat to the earth's biosphere; solar energy devices are called invariant energy systems. Some might add that the fuel is free. But, of course, collecting useful solar energy requires a considerable investment in capital equipment. Until recently, this initial cost for equipment and installation was so high that solar space heating could not compete with cheap fossil fuels. Another attractive feature for some is that for low-technology applications such as for space and water heating, the individual is in control of his or her own energy source. This is of particular importance to those who wish to construct, install, and maintain their own systems.

It is worth emphasizing that solar energy is technically feasible right now. Space heating and cooling systems are being installed at an increasing rate as the economic situation for solar systems improves each

FIGURE 1-7   Forest of mirrors. Faces the 61-meter "power tower" at the Department of Energy's Solar Thermal Test Facility located at Albuquerque, New Mexico. This 5-MeW facility went into operation during the summer of 1978. (Courtesy of Sandia Laboratories, ALbuquerque, New Mexico.)

year. The case for solar electricity is not as well established, since developmental work is just beginning. Initial tests with pilot systems indicate that a net energy gain will easily be obtained. One of the more promising possibilities is illustrated in Figure 1-7. Of course, much engineering development remains to be done before the real economic cost of solar electricity can be properly evaluated. Nonetheless, this is in sharp contrast to the situation with energy from nuclear fusion, which is far from being technically feasible at this point in time. Also, serious environmental problems exist with other energy alternatives. Nuclear fission faces problems associated with reactor safety, routine emissions of radioactivity, the disposal of radioactive wastes, and nuclear proliferation. The large-scale use of coal as an energy source is also fraught with environmental difficulties. According to a recent Stanford Research Institute report, the national program to use coal as the dominant energy source by the year 2000 will require a relaxation of the federal air quality standards, will cause a great change in the land-use patterns in the West, and will require the importation of large amounts of water to areas such as the Colorado River Basin.

BIBLIOGRAPHY

1. Wilson Clark, *Energy for Survival* (New York: Anchor Press/Doubleday, 1975).

    A detailed and clear presentation of our energy problems, environmental concerns, and alternative energy sources. The skillful, interesting prose makes this book both pleasurable and informative. The first three chapters are most relevant to the material discussed here.

2. C. Ruedisili, and Morris W. Firebaugh, eds., *Perspectives on Energy* (New York: Oxford University Press, 1975).

    Contains an excellent series of essays written by experts. Probably the best single source of information on the topic of energy for the nonspecialist.

3. Jack M. Hollander, and Melvin K. Simmons, eds., *Annual Review of Energy,* Vol. 1 (Palo Alto, Calif.: Annual Reviews, Inc., 1976).

    A selection of scholarly reviews of the literature in important areas of energy research. The interested reader should pay careful attention to the articles by Harrison Brown ("Energy in Our Future") and Lee Schipper ("Raising the Productivity of Energy Utilization").

4. "The Biosphere," *Scientific American,* 1-11 (September 1972).

5. "The Potential for Efficient Energy Use," *Science,* 1079 (December 1972); "Efficiency of Energy Use in the U.S.," *Science,* 1299 (March 1973); C. A. Berg, "Energy Conservation through Effective Utilization," *Science,* 128 (July 1973).

6. E. Cook, "The Flow of Energy in an Industrial Society," *Scientific American,* 134 (September 1971).

    This is one of many useful articles in the 1971 Energy issue of *Scientific American.* The reader may also find the article by C. M. Summers (p. 148) on the conversion of energy to be useful reading in this context.

PROBLEMS

1. With 6% of the world's population, the United States presently has the following percentage of the world's energy consumption.
    (a) 50%   (c) 15%
    (b) 30%   (d) 6%
2. Suppose the ambient temperature outside your house is 7°C on a

typical wintry day. Approximately what fraction of your heating needs can be saved by turning down the thermostats from 23°C to 18°C?

(a) 1/6   (c) 2/3

(b) 3/4   (d) 1/3

3. The least efficient mode of freight transport is the:

(a) Railroad   (c) Truck

(b) Airplane   (d) Waterway

4. The total amount of energy currently consumed in the United States per year is about:

(a) 7×10^{16} J   (c) 70×10^{19} J

(b) 7×10^{19} J   (d) 7×10^{12} J

5. The world's consumption of oil will peak near the year:

(a) 1980   (c) 2030

(b) 2000   (d) 2100

6. Suppose that 1000 solar homes are built in the United States this year and that construction grows exponentially, with a doubling time of two years. At what time in the future will the rate of construction be approximately equal to 1 million solar homes per year?

(a) 500 years   (c) 20 years

(b) 32 years   (d) 8 years

7. In the past 30 years the total energy consumption in the United States has been growing at a rate:

(a) Approximately equal to the population growth rate.

(b) Faster than the population growth rate.

(c) Slower than the population growth rate.

(d) No relationship.

8. At present rates of consumption our fossil fuels will soon be depleted. How much faster than their rate of production are we consuming these nonrenewable energy sources?

(a) Twice as fast.   (c) 1,000 times.

(b) 10 times.   (d) 1,000,000 times.

9. Which of the following energy resources is in greatest danger of exhaustion: (Make an educated guess based on your present knowledge.)

(a) Coal.   (c) Shale rock.

(b) Domestic oil.   (d) Arab oil reserves.

10. In the year 2005, American total energy demand will be: (Assume no serious steps toward serious energy conservation are taken.)

(a) The same as today.   (c) Four times as much as today.

(b) Twice as much as today.   (d) Ten times as much as today.

11. The cost of electricity used in producing our common manufacturing items is what percentage of the total production cost?

(a) 95%   (c) 10%

(b) 50%   (d) 1%

12. Which of the following is meant by the expression "energy trap"?
 (a) That we are used to the availability of cheap energy.
 (b) That we are used to the idea of energy being expensive.
 (c) That much of our available energy resources are trapped in a way that renders them useless for practical development.
 (d) That most of our energy resources are trapped in such a way that a very large amount of energy must be expended to release them.

13. As a primary source of energy, coal currently ranks:
 (a) First.   (c) Third.
 (b) Second.   (d) Fourth.

14. The utilization of a finite energy source such as oil follows the familiar "bell-shaped" curve (see Figure 1-6 for oil). The area under this curve represents the total available resource. If the total area under the curve for our oil reserves is increased by 50%, what can you say about the corresponding change in the year at which peak production occurs?
 tion occurs?
 (a) It will change by only a few years.
 (b) It will change by at least 50 years.
 (c) It will occur at a time that is 50% greater than that indicated on the figure above.
 (d) It will not change at all.

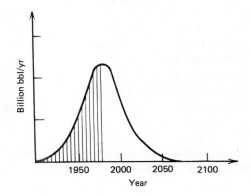

15. Compare the growth in the use of energy in the United States over the last 20 years with the increase in population for the same period.
 (a) The rate of increase of both quantities is the same.
 (b) Usage of energy has increased faster than population.
 (c) Population has increased at a faster rate than energy usage.
 (d) The population has remained constant while the consumption of our energy resources has increased.

16. What is the important concept contained in Harrison Brown's statement that we are caught in an "energy trap"?
    (a) That at present we have an overabundance of energy resources.
    (b) That at present we do not have sufficient energy resources.
    (c) That we are accustomed to energy costs that are much too high.
    (d) That we are accustomed to cheap energy.

17. If the thickness of the insulation in the walls of a building are doubled what can be said about the rate of energy loss through them? (for the same temperature difference between the interior and the outside world).
    (a) It doubles.
    (b) It remains the same.
    (c) It decreases by a factor of 2.
    (d) It decreases by a factor proportioned to $1/\Delta T^4$, where ΔT is the temperature difference.

18. We are going to compare the electric bills during the month of December for two different all-electric houses. House number 1 operate with twenty 100-watt lights, which are each kept on for an average of 10 hours per day. House number 2 has ten 100-watt lights, which are each kept on for an average of 10 hours per day. Compare the electric bill for the two houses at the end of the month.
    (a) The electric bill for house 1 is considerably greater than that for house 2.
    (b) The electric bills for each house are about the same.
    (c) The electric bill for house 1 is considerably less than that for house 2.
    (d) The electric bill for house 1 is twice that for house 2.

19. What is the approximate doubling period for growth since 1920 of the U.S. population? For total U.S. energy consumption?

20. Suppose that 10% of the world supply of gold has been mined, in all of the earth's history up to now. If the doubling period for production of gold is 20 years, how many years will elapse before all of the gold has been mined?

21. What do you think is meant in "Star Trek" by travel at "Warp 9"? Elaborate upon the difficulties of space travel. By making a list of difficulties that must be overcome, speculate upon possible solutions to these space travel problems.

22. Some of the following are possible causes of the energy crisis and some are not. For each decide whether or not it is a possible cause and explain your line of reasoning.
    (a) Exponential growth in a finite system.
    (b) Energy may change form, but cannot be created or destroyed.
    (c) Our reliance upon finite, nonrenewable sources.

(d) Only a small percentage of the U.S. coal reserves are available for strip-mining.

(e) Our explicit reliance upon the low cost of energy.

(f) The recent environmental protection laws and regulations.

(g) Our present inability to develop a working fusion reactor.

(h) The diffuse character of solar radiation implying the need for large area collectors in order to obtain useful energy.

23. Next to each number write a letter giving the most reasonable matching answer. (Sometimes this may require an educated guess.)

____1.	Time for almost total consumption of the world's oil reserves.	(a)	Solar energy per square foot.
		(b)	10 billion years.
		(c)	100 million years.
____2.	Proven U.S. oil reserves.	(d)	1 million years.
		(e)	Stored solar energy.
____3.	Time of formation of fossil fuels.	(f)	10,000 years.
____4.	Age of universe.	(g)	150 years.
____5.	Fossil fuels.	(h)	Proven reserves.
____6.	Photosynthesis.	(i)	10 years.

24. Describe the relevance of each of the following with respect to the useful collection of solar energy.

(a) Renewable energy source.

(b) An invariant energy source (with respect to the earth's biosphere).

(c) Need for storage of energy.

(d) Large area collectors.

25. Devise a method whereby you can estimate the amount of energy used for space heating in an electrically heated house. (*Hint*. There may be some times of the year when no heating is required.)

26. One possible method suggested for energy conservation in residences is to reduce the number of lights turned on. Discuss whether or not this course of action will result in a real energy savings during the winter season.

27. It is pointed out in the text that one way to effect energy savings with cars is to drastically reduce their weight. Suggest at least one other course of action whereby fuel consumption by cars might be drastically reduced.

28. What advantages can you see in having to pay a deposit on bottles? Disadvantages?

29. Give at least two reasons why utilizing returnable bottles for common beverages would be better than using returnable aluminum cans.

30. It has been suggested that a possible environmental problem associated with solar electric plants is that the "apparent" absorptivity of the sunlight incident upon the effective plant area will be

drastically increased. Assume that in fact the actual solar collector area is black (100% absorptance) as compared with only about 30% absorptance for the surrounding area. (1) Show qualitatively that if the overall efficiency of electrical generation is 10%, only a minor perturbation of the local heat balance is made. (2) Devise a way by which even this local inbalance could be corrected.

31. If the thickness of insulation in the walls of a building is increased by a factor of 2, the heat loss through the walls will be reduced by almost a factor of 2. Suppose that the walls of a building are constructed of 4 cm. of concrete. Discuss why 4 cm. of concrete is not nearly as effective in reducing heat losses as 4 cm. of fiberglass insulation. On the other hand, the concrete has the ability to store a large amount of heat as compared to the fiberglass. Discuss how this could be used to advantage in house design.

32. Glass windows on the south end of a residence are one good way to gain useful solar energy for space heating. However, care must be taken in using this approach. Tabulate and evaluate the possible disadvantages of this method of solar heating.

33. Doubling the amount of insulation in the attic of a house will effect a reduction in the upward heat loss of about one-half during the winter heating season. Will this extra insulation in the attic keep the house hotter or cooler in the summer? Explain.

34. One controversial subject at present is whether or not increased energy production is necessary to maintain a healthy economy. For example, one aspect concerns the number of jobs created in constructing a large electric generating plant versus the jobs created in constructing, installing, and maintaining the flat-plate solar collectors needed to produce the useful energy equivalent to that of the power plant. Research this topic and write a short paper. (*Suggestion.* In addition to the entries listed in the bibliography at the end of the chapter, you will find some useful material in the 1977 issues of *Solar Age* magazine.)

// Energy

Energy comes in many different forms. In our everyday experience we see direct evidence of many of these different forms of energy. When we burn wood in a fireplace we are using the stored chemical energy of this fuel. The television waves captured by our TV set represent energy in the form of electromagnetic radiation. A rapidly moving automobile possesses an energy of motion that we call kinetic energy. The advent of the atomic bomb introduced us to the idea of nuclear energy, which is obtained by converting matter to energy. Other common energy terms with which we are familiar are work and heat. Precisely speaking, these last two are not forms of energy; instead they represent measures of the transfer of energy from one form to another. Many of our ideas about energy, such as our understanding of electromagnetic radiation and of the equivalence of mass and energy through the relation $E = MC^2$ (C is the velocity of light) are relatively new. Nevertheless, the concept of energy has proven itself to be both useful and durable.

Energy is an important concept because, so far as is presently known, its conservation is a universally valid law of nature. Without this law of nature, the disappearance of a quantity of energy in one place could not be related to the appearance of a similar amount of energy somewhere else. If it were not for the law of conservation of energy, two different manifestations of energy would involve two different concepts. This concept is schematically illustrated in Figure 2-1, where the stored energy in gasoline is converted into useful motional energy of a car along with a certain amount of nonuseful waste heat due to various inefficiencies in the process of transformation.

In the first chapter conservation of energy was discussed as a possible strategy to help soften the deleterious effects of the growing energy crisis. In that context, conservation implied a reduction in the con-

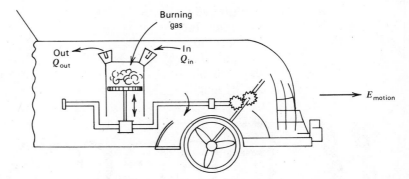

FIGURE 2-1   The stored chemical energy Q_{in} of the gasoline is converted into a variety of other forms by means of the car's
engine. Some of this stored energy goes into bringing
the car into motion, in the case when it is accelerated to
a new velocity. The remainder is either dumped as waste
heat through the exhaust (Q_{out}), or is dissipated as heat
through various frictional processes. Conservation of energy tells us that $Q_{in} = Q_{out} + E_{motion} + E_{friction}$.

sumption of our energy resources. Now we are using the word conservation in a different sense. If a quantity is conserved, this means that
the value or amount of this quantity does not change no matter what
physical processes take place. Used in this way conservation of energy
implies that while the energy content of an isolated system may change
its form, the total amount measured in some appropriate unit remains
constant.

Although energy is a very important concept, it is difficult to state a
precise definition. Energy is not an object that one can see or feel, but
a concept that helps describe the state of an object or a system. For
our purposes a useful definition of energy is that it is the capacity to do
work. In a later chapter this statement will be qualified; appropriate
conversion devices must be available, and the temperature of the surroundings must be sufficiently low. For example, by the use of properly
designed engines, the stored chemical energy in gasoline can be converted into useful mechanical work. It is up to human beings to find
ways of converting the potential capability of our energy sources into
useful work for society in an efficient, economic, and environmentally
clean manner. For most energy sources these last two requirements are
not always compatible.

**A.
SYSTEMS OF
UNITS**

A quantitative discussion of energy requires the specification of a system of units. Intuitively we recognize that units should be specified for the three fundamental dimensions of length, mass (roughly speaking, the weight of an object), and time. Once these three are specified all other mechanical quantities can be expressed in terms of some combination of length, mass, and time. As other physical concepts such as electricity are introduced, one might suppose that fundamental dimensions (such as electrical charge) might be needed. Instead, it turns out the above three dimensions are sufficient to quantitatively describe all known physical quantities. In addition, it is common practice to use a fourth quantity, temperature, as another basic dimension rather than defining it in terms of mass, length, and time.

In terms of the definitions for those fundamental dimensions there are three basic systems of units. These are the MKS system, in which the meter (m), kilogram (kg), and second (sec) are the fundamental units; the CGS system, in which the centimeter (cm), gram (g), and second are the the basic units; and the English system, comprised of the foot (ft), slug [or pound (lb)], and second. The choice of which system to use depends upon the application. The MKS system is advantageous when considering electrical quantities that are expressed in volts (V) and amperes (A). For atomic and nuclear phenomena the CGS system is commonly used. Because of historical precedent, most common quantities are still expressed in English units, although this is now changing as the United States moves toward complete use of the metric system. Metric system units will be adopted in this text, with English units sometimes appended for comparison.

A proper definition of mass requires a study of dynamics, which is beyond the scope of this book. However, an operational definition can be made in terms of a standard block of material. The international standard is a cylinder of platinum-iridium that is defined as 1 kilogram (kg) or 1000 grams (g). A beam balance or similar device can then be used to compare the standard mass with any other object. Secondary standards of greater and smaller mass can be constructed by comparison with the standard mass. In the English system the weight of an object is expressed in pounds. One kilogram is equivalent to the weight of about 2.2 lb of material. The mass of an object differs in a fundamental way from length and time in that the amount of material in an object comes in discrete amounts. All matter is made up of the fundamental constituents of an atom—electrons, protons, and neutrons. Length and time, on the other hand, are infinitely divisible or continuous.

The metric system defines length in terms of a standard meter. This was officially defined as the distance between two parallel scribe marks on a bar of platinum-iridium maintained at a temperature of 0°C. This standard meter is housed in the International Bureau of Weights and

Measures, near Paris, and secondary standards have been distributed to national standards agencies throughout the world. As a way of improving upon the accuracy of the standard meter it was agreed upon in 1961 to define it in terms of a natural unit based on atomic radiation. The meter is now defined as 1,650,763.73 wavelengths of the orange light from Krypton gas, a definition that has the tremendous advantage of being precisely reproducible everywhere. This atomic standard is 100 times as precise as that obtained from measuring the distance between scribe marks on a meter bar. In English units the meter is about 39.37 in.

Webster's defines time as the period during which an action or process continues. Thus, the concept of time is related to the idea of periodic repetition. The essential requirement for a time measurement is to use a process that repeats in a regular, precise pattern. One choice for a standard of time is the mean solar day, the average time interval between successive passages of the sun through a given meridian. The second is then defined as 1/86,400 of the mean solar day. To improve the accuracy of time measurements, recourse was made in 1967 to an atomic standard that now permits an accuracy of one part in 10^{12} to be obtained. Of particular importance to us is the measurement of solar radiation at various times of the day at various locations on the earth's surface. This requires that several corrections to local standard time be made so that solar noon will occur when the sun is directly overhead. First a correction must be made to correct for the local longitude, since the sun requires 4.0 minutes to traverse each degree between the standard time longitude and the local longitude. Also, because of the eccentricity of its orbit, the angular velocity of the earth around the sun is not constant. Because of this, the time period between two consecutive transits of the sun across a given meridian varies with the time of year. Combining this finding with another correction, the fact that the plane of the earth's motion does not coincide with that of the celestial equator, results in a total correction that is commonly called the "equation of time" correction. This correction reaches a maximum of about 14 minutes in February and November (although in opposite directions for the two months). Taking these effects into account allows one to convert from local standard time to local solar time.

The last dimension we need to discuss is temperature. There are a large number of ways to define the concept of temperature. A very imprease method is to utilize our senses to define what is meant by the temperature of an object. This approach discriminates between the temperature of boiling water, ice, and a hot flame, but does not provide a useful quantitative measurement. For more precision it is possible to take advantage of a number of properties of matter that vary with temperature, and these can be used to construct thermometers. For example, mercury expands more than glass when heated, so that the length of the liquid column in a mercury-glass thermometer is a measure of the temperature. Another example consists of two strips of different metals that are joined together side-by-side. When a temperature change occurs, the pair of strips bends owing to the different rates of expansion in the two metals. The higher the temperature, the greater the deflection. When cooled, the bimetallic strip bends in the opposite direction.

To use a thermometer it is necessary to establish a scale of temperature. It is customary to utilize the boiling and freezing points of water to define the scale of temperature. If we call these two points 100° and 0°, respectively, we have the Celsius scale of temperature. If we call these two points 212° and 32°, respectively, we have the Fahrenheit scale of temperature. On the Fahrenheit scale there is a 180° difference between the freezing and boiling point of water, whereas the difference on the Celsius scale is 100°. Thus

(2-1)
$$T_F = 32 + \frac{9}{5} T_C$$

where T_F is the temperature in Fahrenheit degrees and T_C the corresponding temperature in degrees Celsius.

While the mercury thermometer and the bimetallic strip are very useful for many practical applications, they suffer two disadvantages. First, a need is often present for a thermometer that can cover a broad range of temperatures. Also, it is often necessary to describe a scale with temperature defined in terms of degrees above absolute zero. This procedure is essential in the calculation of quantities involving solar radiation. As a suitable standard, covering a wide range of temperature, the helium gas thermometer has been found empirically to be quite satisfactory. The pressure of the gas is proportional to the temperature of this gas over a wide range of temperatures. In mathematical terms this is expressed by saying that the temperature is a linear function of pressure (for a constant-volume gas thermometer). This is illustrated quantitatively by the graph shown in Figure 2-2. An important aspect of this relationship is that if care is taken with the experimental determination of the linear curve relating temperature and pressure, then a good value of the temperature of absolute zero can be obtained by extrapolation to zero pressure; a value close to −273°C is obtained by careful experimentation. It is reasonable to define a new temperature scale with −273°C corresponding to 0° and with the same 100° difference between freezing and boiling points of water as for the Celsius scale of temperature. This new temperature scale is called the Kelvin scale of temperature and is particularly valuable for our purposes, since we will find that the crucial physical phenomena associated with solar radiation are related to the temperature expressed on the Kelvin scale (°K).

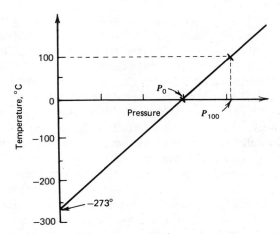

FIGURE 2-2   For a constant-volume gas thermometer there is a linear relation between temperature and pressure over a wide range of these quantities. By measuring several values of p and T, the linear curve can be carefully determined. Extrapolation to zero pressure gives the temperature corresponding to absolute zero.

**B.
WORK, POWER,
AND ENERGY**

Intuitively we all recognize that the concepts of work and energy are related. If we push a lawn mower by hand we must expend a considerable amount of work. By contrast, if a gasoline-powered mower is used, we understand that energy stored in the fuel is converted to useful work in driving the mower. A key feature of the present mechanical age involves the utilization of the energy stored in fossil fuels in such a way that useful work output is obtained. It will be useful to take our present qualitative ideas of work and energy and express them in a careful, quantitative manner.

To precisely define work requires the introduction of the concept of force. In simple terms, force can be defined as the push or pull on an object. Work is then defined as the product of force F by the distance L through which an object is moved. This is expressed mathematically as

(2-2) $$W = FL$$

It is important to note that the weight must move for work to be done. If the weight does not move, no work is done no matter how large a force is exerted. The concept of work is illustrated in Figure 2-3. To move a weight a distance L_1, a force F_1 is required, with an amount of work $W_1 = F_1 L_1$ being done. If the weight of the object is doubled, then

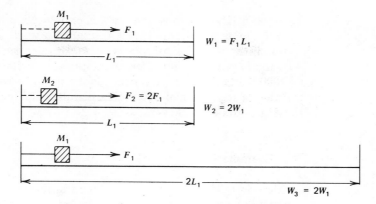

FIGURE 2-3   Illustration of the concept of work and its dependence upon the applied force as well as the distance through which the force is applied.

the resultant force needed is doubled, and the work expended to move a distance L_1 is doubled. If the weight of the object is left unchanged, but the distance moved is changed from L_1 to $2L_1$, then again the work expended is doubled. The metric unit of work is the joule (the work done when a force of 1 newton (N) acts through a distance of 1 meter). The system responsible for the force provides an amount of energy equal to the useful work done—at least for the ideal system discussed here where no other processes take place.

An example of the concepts of work and energy transformations is provided by water spilling over a dam. As the water moves downward it picks up speed. In this case the force of gravity is doing work upon the water molecules. The work done by the gravitational force is converted into motional energy of the water. This motional energy is termed kinetic energy. The change in kinetic energy of the water molecules in going from the top of the dam to the bottom is exactly equal to the work done by gravity. At the bottom the fast-moving water can be utilized to turn the blades of a turbine. The water then slows down and loses kinetic energy, which is in turn converted into mechanical motion of the turbine. The latter can then be used to generate the electricity we so readily use in our modern world.

The rate at which energy is converted from one form to another is often as important as the total energy converted. The term power is technically reserved for this rate. Thus power, P, can be expressed as

(2-3)
$$P = \frac{\text{work}}{\text{time}} = \frac{W}{t}$$

In the MKS system where work is expressed in joules, the unit of power is the joule per second, which is called a watt. The watt is rather small for most industrial purposes; even an electric clock requires several watts of power. A more practical unit is the kilowatt (kW), which equals 1000 watts (W). Large power plants produce useful electrical energy at rates of the order of 1000 million watts, or 1 million kilowatts, or 1000 Megawatts (1000 MW).

Example 1

Electric bills are commonly defined in terms of another practical unit, the kilowatt-hour (kW-hr). Let us express a kilowatt-hour in terms of joules.

$$1 \text{ kW-hr} = 10^3 \text{ W-hr} \times 3600 \text{ sec/hr} = 3.6 \times 10^6 \text{ watt-sec}$$

Since 1 W equals a joule/second, this gives

$$1 \text{ kW-hr} = 3.6 \times 10^6 \text{ J}$$

It would be most logical to use the joule as the energy unit for the many applications involving the transformation of energy from one form to another. Unfortunately, things are not always done in the most logical manner. Before the middle of the nineteenth century the equivalence of the different forms of energy to heat was not understood. For a long time heat was thought to be a fluid, called caloric, which could flow from one object to another. This also resulted in different units for heat than those used to describe mechanical work. The modern view of heat is described in more detail in Chapter X. For present purposes it is sufficient to say that heat is the energy of motion associated with the random movements of the molecules of a substance.

Since heat was originally thought of as a special fluid, it was given its own unit. In the metric system this unit was called the calorie (cal) and was defined as the energy required to raise the temperature of 1 gram of water 1°C. In the English system of units, heat was historically measured in British thermal units (Btu). This unit is defined as the energy required to raise the temperature of 1 lb of water 1°F. Formally, the unit of heat can be related to the change in temperature of a substance by the following equation:

(2-4) $$Q = CM \, \Delta T$$

In this equation M is the mass of the material that has had its temperature changed by ΔT degrees. The constant C is called the specific heat of a substance and is arbitrarily taken to be unity for water in both the English and metric units. It will be seen in Chapters VIII and X that the specific heat of a substance plays a very important role in determining its practical value for the storage of heat. As an example of the use of the relation expressed by Eq. 2-4, the amount of heat required to raise 10 grams of water from 20°C to 35°C is 150 cal. The heat required to raise 10 lb of water from 50°F to 70°F is 200 Btu. The British thermal unit is considerably larger than the calorie, as 1 Btu is equal to 252 cal.

Since heat is only one form of energy, the heat units of British thermal units and calories can be related to work units such as the joule. It was the pioneering work of Count Rumford and Joule that first confirmed the quantitative relation between the two different energy units. Careful experimental measurements established that 1 cal was equal to about 4.18 J. The calorie has now been redefined to be exactly 4.184 J. Energy will be expressed in joules throughout this text. However, since it is still common practice in the United States to express energy in British thermal units, it is useful to be able to convert readily between the two units. Precisely, 1 Btu is equivalent to 1055 J. For most comparisons it suffices to multiply the number of British thermal units by 1000 to get the equivalent number of joules.

Of particular importance to us is the amount of energy contained in the solar radiation coming from the sun. This is usually expressed in terms of the solar energy incident upon a unit area. Commonly this will be in joules per square centimeter (J/cm^2) or in kilojoules per square meter (kJ/m^2). Again, following historical convention, it is customary practice to calibrate solar radiation monitoring equipment in units of Langleys (1), named after an early American solar pioneer. The Langley is equal to an incident energy per unit area of 1 cal/cm^2.

Example 2

Let us calculate crudely the amount of solar energy that annually falls upon the roof of an average house in the United States. The average yearly solar energy incident upon a horizontal surface runs from 300 to 500 l/day in most regions of the country. If we assume that 400 l/day gives

$$\Delta E(\text{year}) = 400 \text{ l/day} \times 365 \text{ days/year} = 146,000 \text{ l}$$

The conversion factor from Langleys to joules per square centimeter is 4.184. Let us also assume a roof area of 1500 square feet. Since 1500 ft^2 = 1.39 × 10^6 cm^2,

$$\Delta E(\text{year}) = 146,000 \times 4.184 \times 1.39 \times 10^6 = 849 \times 10^9 \text{ J}$$

The typical annual space-heating requirements for a house of this size vary from 50 to 100×10^9 J, depending upon climatic region. Thus the solar radiation incident is 8 to 16 times the heating need. Of course, we have not yet taken into account the efficiency of a solar collecting system, the problem of storage, delivery of useful heat to the house, etc.

How does the United States use its energy? The output uses of energy in this country are shown schematically in Figure 2-4. It is instructive to combine the various output uses into specific tasks. If this is done, it is found that space and water heating and air conditioning require almost 25% of the country's consumption of energy resources. Because of this, the application of solar energy techniques for these uses could result in a considerable savings of our nonrenewable energy resources. Another area where solar energy may have a large impact in the future involves the use of process steam by industry, which accounted for 17% of the nation's energy use in 1968. Breaking down the use of energy into specific tasks also confirms the concern about the automobile, which accounted for 13% of the country's consumption of energy resources in 1968.

**C.
ENERGY COST OF
PRODUCING
ENERGY**

A central problem that must be recognized and overcome is that it takes a certain investment of our energy resources to convert the energy contained in other resources into a useful form. For example, to obtain oil from shale rock, the machines that do the needed work of separation consume oil (or the energy equivalent from some other source such as coal). To be profitable, the amount of oil consumed to extract the oil from the shale must be considerably less than that obtained by the separation process.

The failure to recognize the above fact can have very severe consequences. In the midnineteenth century a British company took over the luxury liner Great Eastern, and decided to make use of it for carrying coal from Australia to England. This huge ship, weighing 19,000 tons, and equipped with bunkers capable of holding 12,000 tons of coal, was designed to travel to Australia and back without refueling. Unfortunately, the designers were extremely remiss in their estimates of the coal needs of the ship's engines. It was found that for the Great Eastern to make a round trip to Australia, the ship would need three times more coal than it could carry. The company folded and the ship was broken up for scrap metal in 1888.

The above example illustrates an important lesson. The Great Eastern required more energy to operate than it could carry. The same constraint

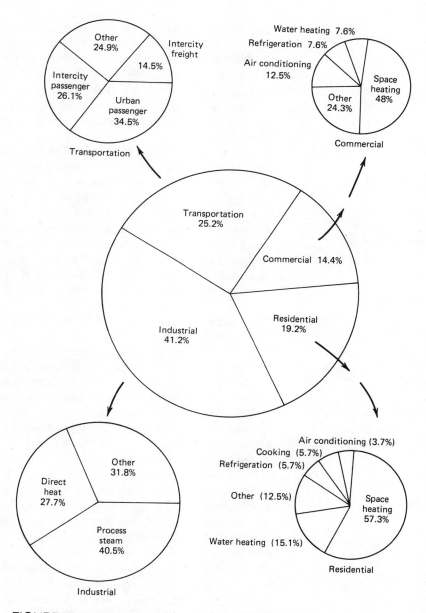

FIGURE 2-4 Graphic representation of the ways the United States uses its energy. (Courtesy of Leo Schipper. From: *Citizens' Energy Workshop Handbook*, Oak Ridge Associated Universities, 1974).

must be imposed upon the future utilization of our energy resources. We have taken considerable advantage in the past of energy sources that were relatively easy to obtain and convert to practical use. However, as more complex energy technologies are implemented, care must be taken to insure that the energy resources consumed are not as large as those obtained. Until recently it appears that this aspect was not seriously considered in the evaluation of potential energy reserves. To better illustrate this point, let us look at a few practical examples.

In 1930 the oil industry found 275 barrels of oil for each foot of exploratory drilling, with the average find at a depth of 3000 ft. Today this figure has fallen to 35 barrels with the average well depth at 6000 ft. Digging exploratory wells to greater depths requires the use of more energy. If we add in the energy "costs" in the future of the problems of offshore drilling and of constructing long pipelines, it becomes apparent that the net gain per barrel of oil produced will be significantly reduced. Despite these increased technological difficulties, the net gain from each unit of oil produced will probably always be a large fraction of the total amount of energy available from the oil obtained in this way.

The same statement cannot be made with such certainty about our other fossil fuels. Strip-mined coal is the easiest to obtain—particularly if proper reclamation procedures are not followed. However, the largest fraction of our coal reserves lies in deeper veins, and will require a significant energy input to mine. There is even greater uncertainty concerning the process of extracting oil from the large potential reserves of shale rock located in the western part of the United States. Oil shale is a sedimentary rock that contains kerogen, a solid, tarlike organic material. This shale rock requires extensive mining, heating, and processing in order to release the oil from the kerogen. It is not certain yet that this extensive working of the shale rock will actually yield a net return in energy, and even if it does the fractional return will probably be low. As another example, the Athabasca tar sands in Central Canada are estimated to contain about 500 billion barrels of oil. However, it requires the consumption of about 90 barrels of oil in order to obtain 100 barrels of oil!

Solar energy applications are in a more favorable position. A solar collector of 100 m² in area will collect enough energy in one to two years to cover the energy resources consumed in the manufacturing process. The situation for most of the large-scale approaches to solar electricity also appears to be equally promising, but a definitive evaluation must await future technological developments.

A detailed assessment of the amount of energy resources consumed in the production process must also take into account the concept of efficiency. No real device, such as an automobile engine, can extract all the energy from a potential energy source and convert it into useful work. Not only are there operating losses such as those due to friction, but even a perfect heat engine has a useful output that is limited by the second law of thermodynamics. The rationale and implications of this law of nature will be discussed in Chapter X. For our present purposes notice that the useful work output from a heat engine such as the gasoline engine in cars on the steam engine in a boat or train is limited by the second law of thermodynamics to be no greater than 30 to 50% of the available energy potential. The actual output will often be considerably less than this. The efficiency η of a device is defined as

$$(2\text{-}5) \qquad \eta = \frac{\text{Useful work output}}{\text{Energy in}} = \frac{W}{Q_{\text{in}}}$$

Thus, the useful work output equals ηQ_{in}. For example, if an oil burning engine has an overall efficiency of 15%, then the net work output from 100 barrels of oil would only be the total energy equivalent of 15 barrels. In the example above of the Athabasca tar sands, it was assumed that the efficiency in using the 90 barrels of oil to obtain the 100 barrels of oil is the same as the efficiency of utilization of the produced oil.

What about the fraction of energy $(1 - \eta)Q_{\text{in}}$ that is not converted into useful work? For example, a nuclear powered electrical generating plant typically has η at about 0.30 (30%). The other 70% of the thermal energy of the reactor is called waste heat and is expelled into the atmosphere, a nearby river or the ocean. The resulting thermal pollution can have serious consequences for the environment. This waste heat does not have to be thrown away in this manner. For example, it could be transported away and used for useful heating purposes. As the cost of energy rises in the future, energy conservation in this manner will undoubtedly become more popular.

The energy ''cost'' of producing electricity with nuclear reactors was the subject of considerable debate for several years. A large input of energy is required for the mining and milling of the natural uranium, to enrich the amount of ^{235}U isotope (see Chapter III) to about 4% from its naturally occurring value of 0.7%, and to construct and operate the reactor. Several analyses of these steps now agree that the electric energy required is about 25% of that obtained from the burning of ^{235}U in an American light-water reactor.

BIBLIOGRAPHY

1. J. Marion, *Physics and the Physical Universe* (New York: John Wiley and Sons, Inc., 1971), Chs. 2 and 7.

2. F. Sears, M. Zemansky, and H. Young, *College Physics* (Reading, Mass.: Addison-Wesley, 1974), Ch. 7.
    The above texts on General Physics give general presentations of the subjects of work and energy. Some mathematical background is needed.

3. G. Crawley, *Energy* (New York: Macmillan Publishing Co., 1975).
    Covers a wide variety of topics on the subject of energy.

PROBLEMS

1. Energy is an important concept in nature because:
   (a) Although its form can change in many ways, it is always conserved.
   (b) It is conveniently related to the square of the force.
   (c) It is simply related to the product of force times velocity.
   (d) The units of energy are the simplest imaginable.

2. A typical unit of energy is the:
   (a) Watt.
   (b) British thermal unit.
   (c) Joules/second.
   (d) Joule-second.

3. Suppose it requires 5.28×10^5 J to lift a weight of 1768 kg from street level to an eighth floor window that is 30.5 meters above street level. How many joules are needed to lift this same weight to the sixteenth floor?
   (a) 2.64×10^5
   (b) 5.28×10^5
   (c) 10.56×10^5
   (d) 21.12×10^5

4. Kinetic energy is:
   (a) Gravitational energy.
   (b) Potential energy.
   (c) Energy of motion.
   (d) Nuclear energy.

5. Energy may be defined as:
   (a) A "push or pull."
   (b) The product of force and velocity.
   (c) The capacity to do work.
   (d) The product of mass and acceleration.

6. Circle all of the following that are equivalent to 1 Btu.
   (a) Approximately 1 kJ.
   (b) 252 cal.

    (c) 1/252 J.

    (d) 3413 kW-hr.

7. Which of the following are units of power?

    (a) Kilowatt-hours.

    (b) Watt.

    (c) Horsepower.

    (d) British thermal unit.

    (e) Kilocalorie.

    (f) Joule/second.

8. Which of the following cannot be defined in terms of a high-precision atomic standard?

    (a) Time.

    (b) Mass.

    (c) Frequency.

    (d) Length.

9. By what factor does the solar radiation incident upon the roof of a typical house during the course of one year exceed its annual space-heating requirement?

    (a) 1/2

    (b) 1

    (c) 3

    (d) 10

10. The concept of "energy cost of producing energy" is important because:

    (a) In order to have a practical energy source, the useful energy output must be greater than the energy input.

    (b) The dollar cost of new resources may be too high to be competitive.

    (c) The net energy gain from the development of new resources is always very large.

    (d) The dollar cost of capital equipment in developing an energy resource is so large.

11. Suppose that a lump of coal with an energy content of 1000 J is burned in a coal-fired electric generating plant. About how many joules of electrical energy will be produced by the plant as a result.

    (a) 800

    (b) 350

    (c) 100

    (d) 50

12. The energy "cost" of a solar collector for space heating will be made up in how many years of operation?

    (a) 20 years

    (b) 10 years

    (c) 5 years

    (d) 1 year

13. Roughly what percentage of the annual consumption of energy resources in the United States goes for space and water heating in buildings?
    (a) 1%
    (b) 5%
    (c) 25%
    (d) 50%

14. The Athabasca tar sands in Canada represent a potential oil reserve of about 500 billion barrels. However, less than 20% of this enormous reserve will be recoverable because of:
    (a) Gravel and sand in the tar beds.
    (b) A very large energy "cost" to recover the oil from the tar sands.
    (c) The necessity of going through a thermodynamic cycle to use the oil produced from the tar sands.
    (d) Conservation measures imposed by the Canadian government.

15. The existence of the law of conservation of energy tells us that:
    (a) Energy cannot be destroyed and therefore can be "recycled" for use as often as needed.
    (b) We need not worry about using up our finite supply of fossil fuels, since the energy is conserved.
    (c) The useful energy output from burning a gallon of gasoline cannot exceed its stored chemical energy content.
    (d) If oil is burned to generate electricity all of the energy of the oil consumed can be converted to useful electricity.

16. What happens to the waste energy from a coal-driven electrical generating plant? Is this energy really lost?

17. Suppose that solar radiation is incident upon a flat surface at the rate of 400 W/m². How much energy is received by 10 m² in one minute?

18. One common way of heating a home is to run electric current through resistive coils located in the floors and ceilings. Suppose that the electric power is generated in a coal-fired plant. Is this a more efficient way to use the energy of the coal for space heating than to directly burn the coal in a furnace? Explain.

19. Explain why you would expect the fuel consumption per mile of travel by a heavy car to be greater than that of a lighter one.

20. Discuss the arguments for and against the use of waste heat from an electrical generating plant for space heating of residences and commercial buildings.

21. Suppose that roughly one small electric clock per person is in use in the United States. Assume also that each clock requires about 4 W of electric power. If all this electricity were generated by one nuclear reactor, what would be its required power level?

22. The students at Henry Elementary School, St. Louis, Missouri, made the following suggestions to reduce our consumption of precious energy resources. Discuss the flaws (if any) in each suggestion.

(a) Lower people's body temperature to 68°F.

(b) Make it a rule that there must be at least two people in every big bed that uses an electric blanket.

(c) Put more hot sauce in the food.

(d) Don't have as many days of school.

(e) Let birds fly around the house in summer to provide better air circulation.

(f) Don't stay in more than one room at a time.

(g) Eat more carrots so that you can see just as good with less light.

Nonsolar Energy Resources

///

One of the formidable technological problems facing this country in the years ahead is to provide an adequate supply of energy without excessive environmental damage. The past 20 to 30 years have been an "oil age" for us, but the limited U.S. supply is already forcing us to import uneconomic quantities of this precious commodity. Clearly this country must move rapidly toward the fuller utilization of other energy resources. The primary aim of this text is to show how solar energy applications can play an important role in providing for future energy needs. In order to do this it is important to assess the status of our other energy resources such as oil, coal, natural gas, shale oil, geothermal energy, and nuclear power.

The literature abounds with widely varying estimates for the energy resources of both the United States and the world. The reasons for this variance can be attributed to both the practical difficulties of estimating reserves and the varying definitions of exactly what should be classified as a reserve. As an example of the latter difficulty, the amount of oil obtainable from oil shale rock could be increased significantly if future technological improvements should permit the utilization of very low-grade shale rock. As an example of the difficulties involved in making reliable estimates, the 1976 U.S. Geological Survey estimates of the reserves of oil and natural gas were revised downward by a factor of 4 to 5 over their 1974 values, mainly as a result of more sophisticated estimation techniques.

Bearing in mind the uncertainties involved in making reliable estimates for the useful energy reserves, let us take a close look at the projected energy resources of the United States. This will provide us with a convenient point of reference for all further discussion. It is useful to make a division of energy sources into those that are depletable (or nonrenewable) and those that are renewable, that is, continually usable. Table 3-1 lists reasonable 1976 estimates of U.S. nonrenewable energy reserves. Noting that the 1976 U.S. annual energy consumption was about 7×10^{19} J, we see that if oil were used to provide for all of the energy needs of the country, there would only be a seven-year supply left. However, it is impossible to extract the oil from the ground at this rapid a rate. In 1977, with oil providing about one-half of the country's energy needs, the amount of imported oil equalled that produced domestically.

One could argue, perhaps correctly, that the oil shale and uranium fusion resources are considerably larger than shown because of the availability of low-grade ores that will some day prove useful. In the case of oil shale, account has not been taken of the energy required to

Table 3-1   Estimated Reserves of Depletable Energy Sources for the United States[a]

Resource	Physical Quantity[b]	Energy Reserve[b] (10^{21} J)
Coal	1.6×10^{12} Tons	39
Petroleum	90×10^{9} Barrels	0.53
Natural gas	490×10^{12} Ft^3	0.53
Oil shale	600×10^{9} Barrels	3.54
Fission (^{235}U)	10^6 Tons (high-quality ores)	0.53
Breeders (^{238}U)	10^6 Tons	74
D-T Fusion	10^6–10^7 Tons of lithium	80–800
Steam and hot water		≈3.8

[a] All estimates are as of 1976. The values taken for oil and natural gas are from the 1976 report of the U.S. Geological Survey (50% probability). The geothermal estimate is from ERDA 76-1, "A National Plan for Energy Research, Development and Demonstration."

[b] These numbers compare with those prepared for the Subcommittee on Energy of the Committee on Science and Astronautics of the U.S. House of Representatives (See references).

Table 3-2 Renewable Energy Resources for the United States[a]

Resource	Amount of Energy (10^{21} J)	Fraction of 1970 Energy Consumption
Solar radiation	46.8	740
Wind power	0.32	5
Ocean thermal gradients	0.38	6
Hydroelectric	0.008	0.14
Photosynthesis	0.015	0.23
Organic wastes	0.006	0.1
Tidal energy	0.006	0.1

[a] C. Starr, Scientific American, *224*, 37 (September 1971).

obtain the oil from the rock. The energy potential of nuclear fusion obtained from the reaction of deuterium with tritium is seen to be quite large, although the amount of basic raw material available is quite uncertain. At present, the extraction of useful energy from fusion processes is still a scientific dream.

Table 3-2 lists the more reliable estimates of the U.S. renewable energy resources, that is, those that are continually available for people's use. The numbers in the last column indicate the fraction of the 1970 U.S. energy consumption that each resource could supply for an indefinite period if each were fully developed. All of these, except for tidal power, are just different facets of solar energy. It is possible that development of photosynthesis in the oceans, such as through the growth of kelp beds, could significantly increase the estimate for this item.

**A.
OIL AND
NATURAL GAS**

Estimates of the amount of oil and natural gas that will ultimately be discovered and produced have differed dramatically over the years. One reason for this is that oil and gas deposits occur in a restricted volume of space and in limited areas in sedimentary basins at all depths from several hundred to several thousand feet, so that reliable yield predictions are difficult. In addition, estimates were often made using improper techniques. In 1956 the renowned geophysicist M. K. Hubbert presented his estimate of the available U.S. reserves, as shown in Figure 3-1, using a sophisticated method of analysis. These early conclusions

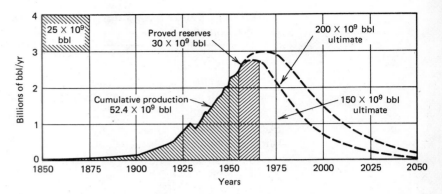

FIGURE 3-1   The U.S. oil production cycle, according to M. K. Hubbert in 1956. Estimates are shown for two different values of the total U.S. oil capacity. [Originally published in the *Canadian Mining and Metallurgical Bulletin*, Vol. *66*, No. 735, pp. 37–54 (1973)].

of Hubbert have since been proved to be remarkably accurate, but at the time were severely criticized. Within a few years after Hubbert's conservative predictions of about 150 to 200 billion barrels of oil as the ultimate U.S. productive capacity, the petroleum industry came up with estimates of the total capacity as high as 600 billion barrels. In 1961 the U.S. Geological Survey presented its estimate of total production as 590 billion barrels of oil. If true, there would have been no expectation of an oil shortage before the year 2000.

How were these higher estimates arrived at? Essentially they followed from the assumption that the amount of oil discoveries per foot of exploratory drillings would remain constant. Therefore, by assuming that only 20% of the oil exploration had been completed in 1959, at which time the total cumulative production was about 100 billion barrels, it follows that the ultimate production capacity would be of the order of 500 billion barrels. The fallacy in this argument is that new oil is becoming much more difficult to find. In the 1930s the oil industry found 275 barrels of oil for each foot of exploratory drilling, while in recent years this has fallen to 35 barrels per foot. An important aspect of Hubbert's bell-shaped production curve, as shown in Figure 3-1, is that changing the total available oil reserve estimate from 150 to 200 billion barrels only changes the year of peak production from 1965 to 1972. In fact, the U.S. production peaked near the year 1970 and has been decreasing since, despite massive efforts to increase production.

In 1975 the U.S. Geological Survey dramatically lowered its estimates of undiscovered, recoverable oil and natural gas. The survey's latest estimates place undiscovered, recoverable oil between 50 and 130 billion barrels. The lower figure is said to have a 95% probability of being right and the higher figure only a 5% chance of being correct. This includes the 10 billion barrels of oil in the Prudhoe Bay Field in Alaska and perhaps 10 to 20 billion barrels of offshore oil. We have arbitrarily chosen a value for inclusion in Table 3-1 that is midway between the two extremes. In view of the 1975 consumption of oil in the United States of about 6 billion barrels, the U.S. reserves appear unlikely to provide significant quantities of oil for more than another 30 to 40 years.

In the face of decreasing production at home, the United States is becoming increasingly dependent upon imported oil, mainly from the Middle East. The principal oil reserves in this portion of the world lie in Saudi Arabia (130 billion barrels) and in Iran and Kuwait (60 billion barrels each). In 1960 two eminent energy authorities were quoted as saying, "Imports involve no foreseeable problem of supply, either in availability at the source or in transportation." Because of the political problems in the Middle East and the 1973 Arab boycott, this position would not appear to be particularly reasonable now.

While the oil situation is serious, our cleanest fuel, natural gas, may be the first to give out. The 1962 estimate of M. K. Hubbert is shown in Figure 3-2, which implies that the U.S. reserves in 1975 would be

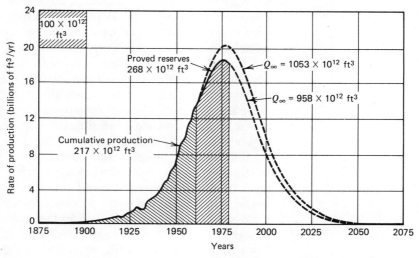

FIGURE 3-2 1962 estimates of M. K. Hubbert of the U. S. natural gas reserves. His prediction of the amount remaining after 1975 is in remarkable agreement with the U.S. Geological Survey estimate. [Originally published in the *Canadian Mining and Metallurgical Bulletin*, Vol. *66*, No. 735, pp. 37–54 (1973)].

about 400 to 500 trillion ft³. This estimate is in remarkable agreement with the latest U.S. Geological Survey estimates of 320 to 655 trillion ft³, with the same confidence percentages for the low and high estimates as for oil. Since 1968, the United States has been using natural gas at a rate that is twice that of new discoveries. Many localities will no longer service new consumers. Imports of natural gas using large, refrigerated tankers that cool the gas to −260°F and reduce its volume by a factor of 600 have helped to temporarily alleviate the shortage.

From Table 3-1, it is seen that shale rock contains large amounts of oil. The Department of the Interior has estimated that the Green River Formation in Colorado, Utah, and Wyoming, shown in Figure 3-3, contains about 600 billion barrels of oil—60 times the Alaskan reserves and about 12 times as much as the established conventional oil deposits in the country. Actually it has been estimated that the Green River shale deposits contain up to 2 trillion barrels of oil, but only about one-third of this is in reasonably thick deposits that average more than 25 gallons of oil per ton of shale; only these are presently regarded as commercially exploitable.

Extracting the oil from oil shale rock is not a simple process, even though the technological details of the extraction process are straightforward. When shale is crushed and heated to 480°C, raw shale oil is released. But, this process is energy intensive, and the energy cost of producing this oil from shale rock must be carefully evaluated. Furthermore, large amounts of water are needed; the production of 1 million barrels a day requires about 3 million barrels of water although recent technological developments may drastically reduce this requirement. By a bitter twist, the shale is located in some of the driest areas in the United States. The primary water supply is the Colorado river and every drop taken increases its salinity.

An exciting development is the possibility of converting the kerogen in the oil shale into crude oil in situ. If successful this technology not only would reduce the cost of shale oil but also would greatly reduce the amount of water required and would alleviate the problem of disposing of the spent shale. A number of approaches of in situ processing have been tried. Figure 3-4 illustrates the scheme being evaluated by Occidental Petroleum Company. In this approach about 20% of the shale rock is excavated and relocated on the surface by conventional means. The remainder of the shale is then fractured with explosives, and the shale rubble expands to fill the voids. The shale is then ignited and air and steam are injected at the ceiling. The oil from the shale flows downward ahead of the flame front and collects in a sump.

B. COAL

Estimation of the amount of coal reserves in the United States and the world can be done much more reliably than was the case for oil. This is because coal is found in stratified beds or seams that are continuous

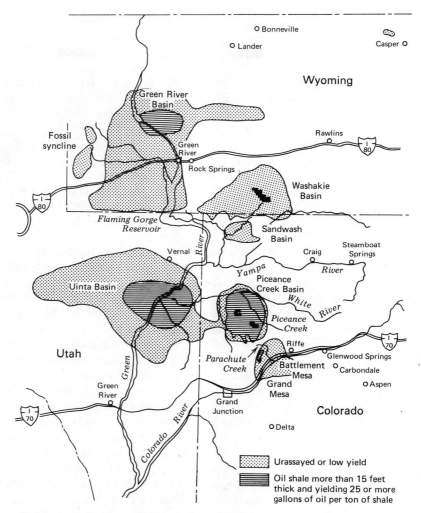

FIGURE 3-3 Map of the major U.S. oil shale deposits. Almost all of the important reserves that are probably recoverable occur in this Green River area. [From: *Science, 184*, 1272 (June 1974).]

over extensive areas, and geological mapping can be done on the basis of a few widely spaced drill holes. On this basis, U.S. reserves are estimated to lie between 1.0 to 2.0 × 10^{12} tons, which is about one-fifth to one-third of the world's supply of recoverable, bituminous coal and lignite. This amount is about 3000 times the 600 million tons of coal consumed in 1970. Our realizable coal reserves may be considerably smaller than this, however, since only a fraction of the total resource

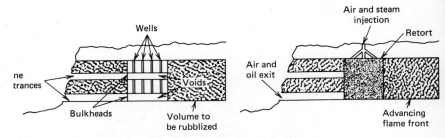

FIGURE 3-4 One suggested approach for the extraction of oil from shale rock in situ. A description of the process is given in the text. [From: *Science, 198*, 1203 (December 1977).]

can be considered minable under present economic and technological conditions. It may be that only 400 to 600 billion tons of coal should be considered as a proven reserve for the United States.

Coal production today is about equally divided between underground and surface mining. An ever increasing amount of coal is obtained by strip mining, even though only about 3 to 5% of our coal can be mined in this manner. This large increase in strip mining has resulted primarily from the introduction of highly efficient, heavy excavating equipment. Mammoth drag lines with bucket capacities up to 168 m², power shovels, and bucket-wheel excavators weighing nearly 8000 tons are now in use. This method enjoys a real cost advantage over deep-mined coal, but so far has left the country with many examples of lunar landscapes, such as that shown in Figure 3-5. Strip-mined areas can usually be reclaimed, and the cost is not prohibitive, but tougher reclamation laws will be needed.

Underground mining of coal also presents many special problems. Dangerous elements such as lung-blackening dusts and explosive methane gas make underground coal mining a risky profession at best. The environmental impacts from deep mining—acid mine drainage, land surface subsidence, gas disposal, and mine and waste heap fires—are far from trivial. Underground mining is generally more wasteful of the coal resource than surface mining; only 57% of the minable coal is presently recovered with the rest of the original resource left in the mine, usually rendered unfit for future recovery.

A serious drawback of coal is its sulfur content, resulting in a large amount of sulfur dioxide (SO_2) after burning. This is commonly removed from the stack gases after combustion. But, a dependable method has not yet been found to accomplish this economically. Several once promising techniques to remove the pollutants from stack gases have had to be abandoned and others are running into trouble. Research scientists

FIGURE 3-5 Typical lunar-appearing landscape from strip mining in North Dakota. This unreclaimed land was mined years ago. Volunteer native shrubs and trees have established themselves in the valleys, and grass sown by the wind covers the windward side of the slopes (the lefthand side of the hills in the picture). No company leaves land in this condition today without violating both state and federal laws. (Courtesy of National Coal Association.)

and engineers are now looking at schemes where the pollutants are removed at an earlier stage in the combustion process. Historically, such a process was used near the turn of the century to provide clean heat at high temperatures for such purposes as heat-treating metals or producing ceramics and fine glassware where the dirty products of coal combustion could not be used. A two-step combustion process was devised. Instead of supplying air to a shallow bed of coal and burning the coal to convert its carbon and hydrogen into carbon dioxide (CO_2) and water vapor, air and steam were blown through a deep bed of coals to obtain a fuel gas composed mainly of carbon monoxide (CO), hydrogen (H), and nitrogen (N). This gas, after cooling and scrubbing with water to remove dust, was burned to provide the desired clean heat. For electricity production, this clean "power gas" can be burned in a gas turbine. Technological improvements in gas-turbine design over the years has resulted in steady increases in both efficiency and size; 300-MW generators are now contemplated. The hot gases rejected from the gas turbine can also be used to drive a conventional steam turbine, with overall efficiencies of over 50% in prospect for the combined two-stage, gas-steam generator system.

Considerations such as the above, combined with the need to use more coal in the future, furnish a considerable economic incentive for coal gassification schemes. One such system is already in commercial operation, the coal-gas process of the Lurgi Gesellschaft für Mineral-öltechnik Company in West Germany. A schematic diagram of this process is presented in Figure 3-6. This system is imperfect, however, since it results in an undesirable release of water vapor, rapid cooling of the power gas, an inability to work on small pellets of coal, and a limited coal-processing capability. Because of the importance of coal gassification, it can be expected that significant improvements will be made in the next few years.

C. AIR POLLUTION

An increase in the use of materials and energy in the United States has resulted in an ever rising rate of pollution. The results of this pollution have now reached the point where the adverse environmental effects are noticeable on a large scale. David Freeman, formerly Chief of the Energy Policy Staff at the White House, has stated the situation quite clearly and succinctly, "Americans as a nation no longer feel that we can produce and use energy with total disregard for its effect on the environment."

The many aspects of pollution and its control form the basis of entire books. All that will be covered in this section is a quick survey of some of the key pollution problems associated with the use of fossil fuels. As noted earlier, one serious problem is the resultant water and ground contamination from oil and coal production. Another serious aspect is the air pollution resulting from the consumption of fossil fuels to perform useful work. We are perhaps most acquainted with air pollution through the unhappy phenomenon of smog, a collection of irritating hydrocarbons and oxides of sulfur and nitrogen. The dominant air pollutants in the United States are listed in Table 3-3.

Since the burning of coal, particularly to produce electricity, is liable to be the dominant source of energy in the United States over the next century, it is worthwhile to emphasize the serious nature of the environmental concerns that result. In a large coal burning power plant the principal emission from the plant's exhaust stacks is carbon dioxide (not shown in Table 3-3). Although carbon dioxide is not a serious pollutant in itself, there is growing concern that the dumping of vast amounts into the earth's atmosphere could seriously affect the planet's climate. The most harmful pollutants released are oxides of sulfur. Large emissions of these sulfur compounds can cause deaths from cancer, damage to human lungs, and property damage. Other dangerous

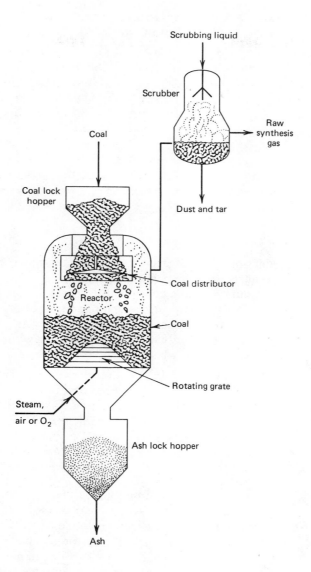

FIGURE 3-6 Schematic drawing of the Lurgi coal gassification scheme. This process uses a rotating grate underneath the coal bed for feeding oxygen and steam. [From: H. Perry in *Perspectives on Energy*, C. Ruedisili and M. W. Firebaugh, eds., Oxford Press, New York, 1975, p. 399. Originally published in *Chemical Engineering*, Vol. *81*, No. 15, 88–102 (1974).]

Table 3-3   Estimated Emissions of Air Pollutants by Weight in 1969[a]. Total Emission was 281.2 Million Tons.

Pollutant	CO	Particulate	SO	HC	NO	Total
Auto	111.5	0.8	1.1	19.8	11.2	144.4
Power plants	1.8	7.2	24.4	0.9	10.0	44.3
Industrial	12.0	14.4	7.5	5.5	0.2	39.6
Refuse burning	7.9	1.4	0.2	2.0	0.4	11.9
Miscellaneous	18.2	11.4	0.2	9.2	2.0	41.0
Total	151.4	35.2	33.4	37.4	23.8	281.2

[a] Units are millions of tons per year.

emissions are nitrogen oxides, the principal pollutants in automobile exhausts and fine particulate matter.

Pollution can be controlled. Emissions from the 950-MW steam plant at Bull Run near Oak Ridge, Tennessee, are relatively low. A giant bank of precipitators traps most of the fly ash, about 950 tons/day. This can be contrasted with plants near Chicago and the Four Corners area of New Mexico, where the fly ash pours out, although recent efforts in New Mexico have resulted in an order of magnitude improvement.

Not all pollution control need be performed at the source. One of the better ways to minimize the adverse effects of urban industrialization is through the use of green areas such as golf courses, parks, and so on. It has been found that a park or golf course, plentifully covered with trees, can effect a 10 to 25% reduction in air pollutants such as SO_2, oxides of nitrogen (NO), and hydrogen sulfide (H_2S), in the neighboring area.

Perhaps the most attractive feature of utilizing solar energy is that it is essentially a nonpolluting energy resource. Space heating of homes by solar energy is an ideal example of this. A solar collector system gathers the incoming insolation, stores it, and distributes it to the house, with a resultant leakage to the outside atmosphere. There is practically no change in the heat balance of a city between the situation with and without solar heated buildings. No pollutants are emitted from the house as is the case for gas, oil, wood, and coal furnaces. (For electric heating, the pollutants are emitted at the generating plant.) The energy and economic cost of solar space heating system is not much greater than that of a conventional heating system. The impact upon the environment of solar electric plants is harder to evaluate, since all of these are in very early stages of development; but they should be relatively nonpolluting by comparison with other energy sources.

**D.
GEOTHERMAL
ENERGY**

Geothermal heat comes from hot rock in the earth's interior, which is at temperatures near 500 to 1100°C. As it is conducted to points near the surface, ground water is heated to temperatures near 200 to 400°C. The resulting geothermal energy resources can be divided into four distinct types.

Dry steam wells are formed when the ground water deep within the earth has a high temperature at low pressure resulting in underground boiling. The accumulated steam can be tapped by wells ranging in depth from 600 to 2400 meters, releasing the dry steam directly. The hot steam is then run through centrifugal separators to remove the rock and grit. From these it is fed directly into low-pressure steam turbines to produce electricity. Electricity is now produced from dry geothermal steam fields in Larderello, Italy (384 MW); the Geysers Field, Sonoma County, California (522 MW); and at the Matsukowa Field in Japan (50 MW).

Wet steam geothermal areas are much more abundant than dry steam areas. In this case, hot water in the earth is kept under high pressure, and boiling does not occur underground. When tapped, the superheated water flows into the well and changes into a vapor. The steam must be separated from hot water before it can be used to generate electricity. The waste hot water is allowed to run off, and since it contains many pollutants, serious environmental problems are involved.

Hot water fields at 65 to 120°C are primarily used for heating or refrigeration. They can also be used to generate electricity by boiling a secondary fluid such as freon, with a low boiling point, with the secondary fluid driving power turbines, but the energy conversion efficiency is very low. Small plants of this type are being used in New Zealand, in Kamchoika in the USSR, and in Japan. The city of Reykjavik, Iceland, is heated geothermally, with hot water used for domestic purposes, swimming pools, etc. Using the hot water for heating in this manner may prove the best way to tap the energy of this large geothermal resource.

A new program to tap the geothermal energy contained in hot rocks at depths of more than 4000 meters is being undertaken by the Department of Energy. Pressurized water is used to "fracture" the hot rock, as in oil drilling. Water is brought down to this depth through one pipe where it turns to steam. This steam is then piped up through another hole that has been drilled close by and flows into a turbine generator where electricity is produced. The hot rock fractures continuously and conceivably such a well would last about 30 years, producing power on the order of 50 to 100 MW.

**E.
NUCLEAR FISSION**

Of all the new sources of energy, nuclear fission has received the most support, and its technology is the most well developed. Over 50 nuclear reactors are now operational in the United States, and many more are in the planning stage. Until recently it has been commonly thought that nuclear reactors would rapidly take over from fossil fuels the lead role in the generation of electricity. While this may still come about, there is a growing concern in both the general public and the scientific community about the extent to which research funds have been concentrated on this energy option and about the wisdom of basing an energy economy upon nuclear power. Increasing awareness of the possible consequences of the large-scale use of nuclear fission as a source of energy has also contributed to this reevaluation of our energy priorities. The debate over the pros and cons of nuclear power has often risen to a white-hot fury. The actual determination of a national policy on this subject is still to come. Nevertheless, the nuclear age will undoubtedly be with us for a long time.

The emphasis in this section is upon the potential of conventional nuclear reactors as an energy source. The essential nuclear physics needed to understand the fission process is well covered in the many fine books on this topic. It is sufficient for our present discussion to note that when a slow neutron interacts with a fissionable isotope such as ^{235}U (this is uranium with 92 protons and 143 neutrons. Natural uranium also has an isotope, ^{238}U, with 92 protons and 146 neutrons.), the resultant object will split into two almost equally massive fragments, with a consequent large release of energy. Energy is also released when two light nuclei are combined together in a fusion reactor. However, since the probability of obtaining useful energy from fusion reactions in the next 20 to 30 years is not very large, barring some unforeseen scientific breakthrough, this subject will not be covered here.

A nuclear reactor is based upon the principle that the fission processes occur in an environment such that the number of neutrons given off during fission are just exactly sufficient to maintain the process. This means that if 2.5 fast neutrons are emitted, on the average, during a typical fission event, then exactly 1.0 neutrons must remain, on the average, after the emitted neutrons have undergone a typical life cycle.

Most nuclear reactors in the United States have a number of common features. The fissioning material, ^{235}U, is formed into long slender rods spaced uniformly throughout the core of the reactor. (Most of the uranium in the fuel rods is ^{238}U, since only about 3% of the uranium is ^{235}U.) The neutrons emitted in the fission process come off with very large velocities and cannot cause ^{235}U to fission. This necessitates filling the reactor volume between the fuel rods with a light material called the moderator, which serves the purpose of slowing down the fast neutrons. Because of the fissioning material, the fuel rods become hot and a coolant, usually water, is passed over them in order to remove the energy generated. In order to extract useful power the hot coolant is then passed through some type of turbine where electric power is generated. To maintain the reactor in a stable operating condition and furnish protection in the case of the failure of any of the reactor components, control rods that readily absorb neutrons are needed. Protection against a loss of water is provided by an emergency cooling system.

It is instructive to examine the life cycle of a typical neutron emitted from a fission event. The average history of such a neutron is depicted in Figure 3-7. Some of the fast neutrons created in the fission process either leak out of the reactor volume or are absorbed by the uranium (mostly ^{238}U) in nonproductive processes. This happens during the time period when the moderator is slowing down the fast neutrons. After this slowing-down process, about 1.7 of the original neutrons are left (on the average, of course). Of these slow neutrons, 0.5 are absorbed by material other than ^{235}U, and 0.2 leak from the reactor volume. In the stable situation exactly 1.000 . . . . neutrons are left over to instigate a new fission reaction.

It only takes a small fraction of a second for this neutron cycle to be completed. If all of the neutrons in the reactor went through exactly this cycle, control of the power level of a nuclear reactor would be impossible. Fortunately, about 1% of the emitted neutrons arise from processes that are very much delayed from the original fission event. Because of this fortuitious circumstance, reactors can be controlled by the relatively slow movements of control rods.

Natural uranium is a plentiful element, even being found in small amounts in granite and in sea water. However, only about 0.7% of natural uranium consists of fissionable ^{235}U, the rest consisting of ^{238}U, which is an isotope containing three more neutrons and which is not fissionable by slow neutrons. Because of this circumstance, there is a

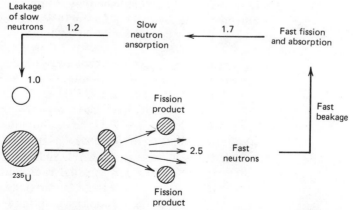

FIGURE 3-7   Schematic illustration of the life cycle of neutrons in a typical reactor situation. At equilibrium, the original 2.5 neutrons are just sufficient to overcome leakage and unwanted absorption processes, providing one slow neutron to initiate another fission event.

lower limit to the concentration of uranium in naturally occurring ores that can usefully be exploited. An estimate of the useful uranium reserves in the United States as published in a 1973 report by the old Atomic Energy Commission is given in Table 3-4.

The reserves are listed in terms of tons of uranium oxide (U_3O_8). For example, the first row shows that there are about 1.13 million tons of uranium oxide available whose naturally occurring concentration is greater than or equal to 1600 parts per million (pure uranium ore). There are an additional 1.63 million tons of ore with a concentration ranging from 1000 to 1600 parts per million, and so on. One sees immediately that the fission reserve for conventional reactors is rather small if only the low-cost, almost pure uranium ore reserves are used. In fact the energy potential from this fraction of the uranium reserve is no greater than that from the remaining U.S. oil reserves. Making use of the lower-grade ores will alleviate this situation somewhat. The real attractiveness of nuclear power for the long term lies with the potential of the breeder reactor. This approach permits more fissionable material to be manufactured than is consumed. This occurs because ^{235}U, the dominant uranium isotope, can be converted into ^{239}Pu when it interacts with a fast neutron. And ^{239}Pu is a fissionable isotope, just like ^{235}U.

A reactor cannot be constructed of natural uranium because the much more plentiful ^{238}U absorbs too many neutrons to permit a chain reaction to take place. However, a chain reaction can occur in a large lattice-type structure composed of uranium metal interspersed with a light-weight moderator such as ^{12}C, water, or "heavy water" (D_2O). The purpose of the moderator is to slow down the fast neutrons as quickly as possible in order to avoid their being captured in nonfission processes. This is accomplished by using the fact that whenever a fast neutron collides with a light atom it loses energy and hence has a smaller

Table 3-4   U.S. Uranium Reserves

U_8O_8 Concentration (parts per million)	Available U_3O_8 (10^3 tons)	Electrical Energy[a] (10^{21} J)	
		Conventional Reactors	Breeder Reactor
1600	1,127	0.21	28.2
1000	1,630	0.30	40.6
200	2,400	0.44	59.5
60	8,400	1.56	208
25	17,400	3.26	431

[a] Includes an overall efficiency of 32% in conversion to electric power.

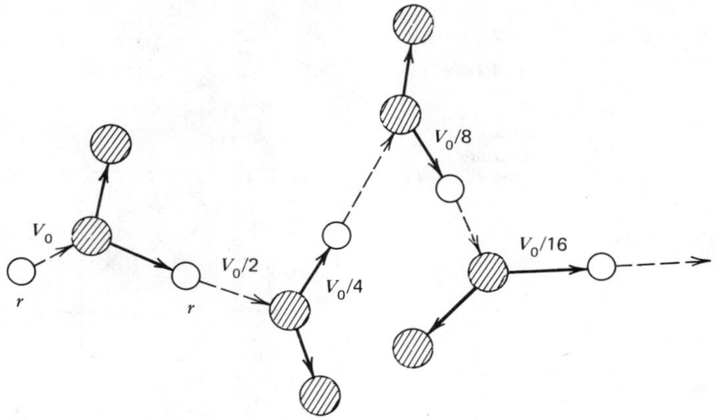

FIGURE 3-8   Graphical illustration of the slowing-down process going on in a nuclear reactor; fastest slowing down occurs when the moderator atoms have the same weight as the neutron, that is, in the case where ordinary light water is the moderator. However, see text description for the disadvantage of light water. In reality the scattered neutron can have a wide range of velocities; the choice of one-half of the initial velocity was made only to schematically indicate the slowing-down process.

velocity; the lighter the struck nucleus the more energy the neutron loses. On the average the maximum energy transfer occurs when the struck nucleus has the same mass as the neutron. A graphical portrayal of the slowing-down process is given in Figure 3-8. From this standpoint the hydrogen in ordinary water would be the best moderator; unfortunately, part of the time hydrogen captures a neutron to form deuterium, which is an additional absorption process that does not lead to useful fission. Consequently, ordinary water cannot be used with natural uranium as the fuel, but instead it requires uranium that has been enriched in the content of ^{235}U.

For reasons that were partly economic and partly political, industry in the United States has chosen to go with ordinary light water as the coolant and moderator. This water is circulated under pressure thereby allowing elevated temperatures up to 330°C to be achieved. There are two types of light-water reactor that are now in use. In one, the pressurized reactor, used in about 60% of the nuclear power installations, the coolant is passed through an apparatus that heats water in a separate, closed cycle, and steam from this system is then channeled to a turbine driving an electric generator. A schematic diagram of a pressurized-

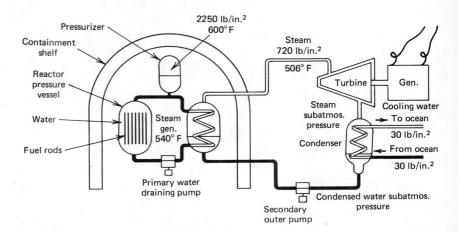

FIGURE 3-9 Illustration of the basic components of a conventional pressurized, water fission reactor. (From: *Perspectives on Energy*, C. Ruedisili and M. W. Firebaugh, eds., Oxford University Press, New York, 1975; with the permission of the National Academy of Science, Washington, D.C.)

water reactor is given in Figure 3-9. The other commercial installations are known as boiling-water reactors. In this case the water in the reactor is held at a lower pressure and is allowed to boil inside the reactor core. The steam and water flowing past the fuel are then separated with the steam flowing directly to a turbine. Both types of reactors have thermal efficiencies (electrical power output/energy input) of about 32%, compared to a typical value of 40% for a fossil-fueled plant that can operate at higher temperatures.

Since the U.S. reactors utilize light water as the moderator, they must then enrich the percentage of ^{235}U in naturally occurring uranium. This is presently done at Oak Ridge, Tennessee, where solid U_3O_8 is converted to gaseous uranium hexafluoride (UF_6) and then passed many times through a gaseous diffusion process. The uranium fuel, enriched to 3% ^{235}U is then suitable as the fuel for the light-water reactors. Most of the energy "cost" for the nuclear power industry occurs in this step. In order to enrich the ^{235}U content to 3% for the desirable fuel, 25% of the original ^{235}U is left over as reject material in the form of depleted uranium "tailing," made up of uranium with a ^{235}U content of 0.2%. Because of the need to enrich to 3%, it takes about 5 tons of pure uranium ore to give 1 ton of enriched fuel. A typical 1000-MW reactor uses about 30 tons of enriched fuel in one year. Present reactor practice is to consume the ^{235}U in the reactor until the ^{235}U percentage is down to about 1%. Ordinary water is not the only choice for a moderator, as we have already noted. The original Hanford reactors, designed to produce the fissionable isotope ^{239}Pu during the Second World War, used very pure graphite as the moderator. Heavy water is used as the moderator in the Canadian reactor program. This permits these CANDU reactors (for Canadian-deuterium) to use natural uranium as the fuel. Since no enrichment is needed with its subsequent rejected material, and since almost complete burnup of the ^{235}U is possible in this approach, the CANDU reactors can get about twice as much energy from a ton of natural uranium as does the typical light-water U.S. reactor under present operating conditions, when no reprocessing of the spent fuel is done.

In many ways nuclear fission has substantial advantages over traditional sources of energy. Air pollution from the consumption of fossil fuels is still a problem, while nuclear power plants do not emit particulates, sulfur oxides, or other combustion products. Nuclear fuels are a compact source of energy, which has advantages in terms of size and location of power plants. If the breeder reactor proves to be a valid future energy source, then nuclear power can provide for the world's energy needs for perhaps several thousands of years. Nevertheless, society has not yet generally accepted nuclear power as the best solution to the world's energy problems. The major concerns about nuclear power can be grouped roughly into four categories: (1) routine emission of radioactivity from the reactor, (2) radioactive waste disposal, (3) safety problems such as the possibility of a loss of coolant accident, and (4) the problem of safeguarding an ever increasing inventory of fissionable material. Other problems such as thermal pollution from waste heat are common to all electric power generation schemes.

BIBLIOGRAPHY

1. L. Ruedisili and M. Firebaugh, eds., *Perspectives on Energy* (New York: Oxford University Press, 1975).

   One of the very best of the recent books on energy. This volume is a collection of essays by experts on the basic issues as well as on fossil fuels, nuclear power, and other possible energy alternatives. The article by M. K. Hubbert that surveys the world's energy resources is especially recommended.

2. Robert H. Romer, *Energy, an Introduction to Physics* (San Francisco: W. H. Freeman and Co., 1976).

   Appendices H, I, K, and L of this general physics text contain a wealth of valuable data about the energy content of fuels, about production and consumption of energy, and about our sources and uses of energy.

3. Jack M. Hollander and Melvin K. Simmons, eds., *Annual Review of Energy,* Vol. 1 (Palo Alto, Calif.: Annual Reviews, Inc., 1976).
   Contains a number of useful, readable articles on coal, coal gassification, geothermal energy, and oil shale.

4. ''Energy Facts'' (Prepared for the Subcommittee on Energy of the Committee on Science and Astronautics, U. S. House of Representatives, Ninety-Third Congress, November, 1973; U.S. Government Printing Office, 99-7220).
   A very extensive compilation of data.

5. ''Energy,'' *Science, 184* (April 19, 1974).
   This complete issue of *Science* was devoted to energy and included such diverse topics as the impact of the energy crisis, national policy matters, and solar energy by photosynthesis, as well as articles on the developed energy technologies: oil, coal, gas, and uranium.

6. ''Geological Survey Lowers Its Sights,'' *Science, 189,* 200 (1975).
   Summarizes the latest U.S. Geological Survey estimates of oil and natural gas reserves.

PROBLEMS

1. Which of the following countries has the largest oil reserves (as of 1977)?
   - (a) Saudi Arabia.
   - (b) United States.
   - (c) Norway.
   - (d) Japan.

2. Which of the following U.S. energy resources has the greatest potential energy content?
   - (a) Oil.
   - (b) Coal.
   - (c) Natural gas.
   - (d) Tar sands.

3. The present U.S. oil reserves are sufficient for:
   - (a) 30 to 40 years.
   - (b) 100 years.
   - (c) 1 to 5 years.
   - (d) Forever.

4. The energy resource that may give out soonest is:
   - (a) Conventional nuclear reactors.
   - (b) Natural gas.
   - (c) D-T fusion.
   - (d) Geothermal.

5. The three largest, major, geothermal power plants in the world are located in the United States, Italy, and Japan. They represent power generated from:
   (a) Hot water.
   (b) Hot water plus steam.
   (c) Hot steam.
   (d) Hot water obtained by pumping cool water down into very deep wells.

6. Can sufficient hydroelectric power be generated from U.S. rivers in the future to produce the amount of energy used in the United States in 1970?
   (a) Depends upon an improvement in generator efficiencies.
   (b) Yes, if fuel cells become practical.
   (c) Yes.
   (d) No.

7. The type of geothermal resource that can be least efficiently converted to electricity is:
   (a) Dry steam.
   (b) Wet steam.
   (c) Hot water.
   (d) Hot rocks at a depth of 15,000 ft.

8. The air pollutant that is most dangerous to one's health is:
   (a) Carbon dioxide. (c) Carbon monoxide.
   (b) Particulate matter. (d) Oxides of Nitrogen.

9. It has been suggested that coal will be used to power the automobile of the future. This might reasonably be accomplished by:
   (a) Converting the coal to water vapor.
   (b) Conversion of the coal to power gas.
   (c) Direct burning of the coal in the automobile.
   (d) Making a liquid solution of coal and lime.

10. It is easier to make good estimates of coal reserves than of oil or natural gas because:
   (a) The coal comes in stratified beds or seams.
   (b) The coal reserves are mostly located near the earth's surface.
   (c) The good oil and natural gas reserves are all located more than 5 miles below the earth's surface.
   (d) Most of the nation's coal reserves are located in the Green River area of Utah and Wyoming.

11. What is the moderator in most of the conventional U.S. power reactors?
   (a) Carbon. (c) Heavy water.
   (b) Uranium. (d) Ordinary water.

12. Which of the following reactor accident scenarios is generally considered to be the most serious?
   (a) One in which a control rod sticks inside the reactor.

(b) One in which the uranium fuel sticks inside the hollow tube that encloses it.

(c) One in which a total loss of flow of the normal water supply occurs.

(d) No serious accident can possibly occur.

13. Control of a reactor by means of neutron-absorbing control rods is feasible because:

   (a) A small amount of ^{240}Pu is present in the fuel.

   (b) Not all fission events result in neutrons being emitted.

   (c) Some neutrons from fission events are delayed in their emission.

   (d) Some ^{233}U is mixed in with ^{235}U in the nuclear fuel.

14. The function of the moderator in a reactor is to:

   (a) Provide additional safety.

   (b) Slow down fast neutrons.

   (c) Increase the breeding ratio.

   (d) Provide structural strength.

15. The CANDU reactor uses which of the following as the moderator?

   (a) Heavy water.   (c) Carbon.

   (b) Light water.   (d) Natural uranium.

16. The emergency coolant system for a nuclear reactor is designed to do which of the following tasks?

   (a) To shut down the reactor in case the emergency safety rods fail.

   (b) To cool off the reactor from the energy generated by the residual fission product activity after a shutdown whereby a loss of coolant has occurred.

   (c) To cool the reactor fuel rods in steady-state operation when the regular water system fails.

   (d) None of the above.

17. Review the literature on coal gassification schemes and tabulate the advantages and disadvantages of each.

18. Discuss the global energy balance associated with the use of solar energy to heat buildings. How does it differ in this respect from the use of coal and natural gas?

19. It has been said that coal will be used to fuel automobiles, airplanes, and home furnaces of the future. How is this possible?

20. Shale oil is potentially capable of producing all of the U.S. oil needs for perhaps 100 years, but will probably never provide more than 5 to 10% of our present yearly consumption. Explain.

21. The burning of our limited fossil fuel reserves has been criticized because of the need of future generations to use fossil fuels as raw materials in producing chemicals, synthetic fibers, etc. Name three energy sources for which this criticism would not apply.

22. One production estimate claims that three barrels of water will be needed to make one barrel of oil from shale rock. A reasonable value of the surface water available for this purpose in the Green River Basin is 3 million barrels of water per day. Assuming that the U.S. consumption of oil is 6 billion barrels of oil per year, what fraction of this could be taken up by oil from shale rock.

23. According to recent newspaper accounts, the proven natural gas reserves in the Prudhoe Bay Field are about 30 trillion ft^3. In Chapter I we saw that natural gas presently provides about 30% of our annual energy consumption. For how long would the Alaskan Fields cover our total national natural gas needs? (*Hint*. 1 ft^3 of natural has has an energy content of about 10^6 J.)

24. Would going over to electric cars eliminate the pollution problems associated with this mode of transportation?

25. Suppose that modern pollution control devices on automobiles reduce emissions by a factor of 2 while at the same time gasoline economy is reduced by one-third. What is the real improvement in the amount of auto related pollutants?

26. What is the principal reason why geothermal resources are best suited for direct heating use rather than the generation of electrical power?

27. From a study of Figure 3-1, estimate the year of peak U.S. production of oil if the total (including that already pumped) oil reserve were 250 billion barrels. What does this say about the sensitivity of the peak production year to the total oil capacity.

28. Suppose that we wish to carry out calculations of the energy cost of producing energy for the case of coal. Assume that a unit of coal, which can be burned to produce 100 Btu of heat, can be mined, transported, and so on at an energy cost of 5 Btu of electrical energy. If a coal-burning electrical generating plant has an overall efficiency of 30%, what percentage of the output electrical energy is used to produce it?

29. Suppose that a fast neutron in slowing down by collisions with moderator atoms loses one-half of its energy of motion during each collision. How many collisions are required to slow a neutron down to the point where its energy of motion is only 6% of its initial value?

30. Explain why deep salt bed formations appear to be an attractive location for the storage of radioactive wastes.

Introduc- tion to Solar Energy

IV

A newspaper in France in the seventeenth century contained the following account of the destruction of the mighty Roman fleet under the command of Marcellus, during the siege of Syracuse in the year 212 B.C.

Archimedes set fire to Marcellus' navy by means of a burning glass composed of small square mirrors moving every way upon hinges which when placed in the sun's rays, directed them upon the Roman Fleet so as to reduce it to ashes at the distance of a bowshot.

This account refers to the reputed use of solar energy by the great scientist and mathematician, Archimedes. The veracity of this legend has been disputed by many historians. However, devising such a scheme would have easily been within the powers of Archimedes, who had already distinguished himself by designing a number of innovative war machines, which had played a key role in the defense of Syracuse. If we use a rough estimate of the solar energy incident at this latitude, it can be shown that temperatures of 850 to 1000°C could have been attained by an absorbing object located 30 to 60 m away from properly oriented reflecting mirrors of a thousand square meters in area. This concentration of solar energy would have been more than sufficient to ignite the dark-colored sails of Marcellus' fleet in a few seconds. Better documented by historical records was the use of mirrors by Procleus in 626 A.D. to repel the fleet of Vitellius during the siege of Constantinople.

For the next 1000 years, most of the devices invented to use solar energy were constructed for amusement purposes. An example of this is the solar fountain of De Caus, which is illustrated in Figure 4-1. Several glass lenses were mounted on a frame in order to concentrate the sun's rays on an airtight metal chamber partially filled with water. The sunlight heated the air of this primitive solar engine, forcing water out as in a small fountain.

More serious attempts to make use of the sun's energy soon followed. The French scientist George Buffon constructed the first multiple-mirror solar furnace in 1747. He used an array of 168 flat, square mirrors, each 15 cm on a side, to ignite a woodpile at a distance of 60 m. The famous chemist, Lavoisier, also experimented with solar furnaces because they allowed high temperatures to be obtained in a clean environment. With a special 130-cm diameter lens filled with alcohol to increase its refractive power, he was able to melt platinum at 1780°C. A schematic diagram of this arrangement is shown in Figure 4-2. The French have maintained an interest in solar furnaces ever since, leading up to the large 1-MW facility of F. Trombe now in use at Odeillo, France. The Italians were not quite so prudent in their work in this field. In 1695, Targioni and Averoni decomposed a diamond by concentrating the sun's light on it with a lens. No more startling discoveries came out of the Italians, and it is assumed that their unwise choice dealt them an unsurmountable financial blow.

Pioneering work with flat-plate collectors was done during the middle of the eighteenth century by the Swiss scientist, Nicholas de Saussure. He designed a solar oven consisting of glass plates spaced above a blackened surface enclosed by an insulated box. The sunlight entered the box through the glass and was absorbed by the black surface; by chemically coating the outside surfaces of the glass, he was able to achieve temperatures as high as 150°C. Because of the low density of the incident solar radiation, this simple approach is still the most inexpensive and reliable way to obtain energy in the form of heat from the sun's rays.

A concentrating type of solar cooker was described in an article in *Scientific American* in 1878 by W. Adams of Bombay, India. As shown

FIGURE 4-1   Solar water fountain designed by Soloman de Caus in 1615. [Reprinted with permission from *Sunworld*, No. 5 (August 1977), pp. 17–32, published by Pergamon Press, Ltd.]

in Figure 4-3, this apparatus uses a conical, octagonal box lined with silvered glass mirrors to focus light through a cylindrical bell jar onto the food container. This invention worked very well. Another of his suggestions for using solar heat—for the cremation of deceased Hindus and others—was not well received. A solar cooker that not only employed parabolic mirrors to reach high temperatures but that also used

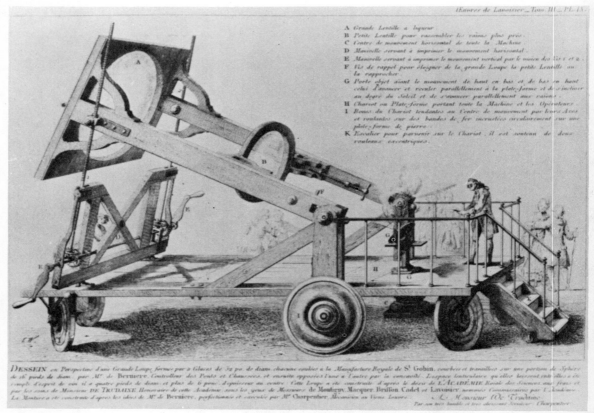

FIGURE 4-2 The burning apparatus of Lavoisier. [Reprinted with permission from *Sunworld*, No. 5 (August 1977), pp. 17–32, published by Pergamon Press, Ltd.]

a means of heat storage enabling food to be cooked after sundown was developed by Dr. Charles Abbott, the dean of American solar scientists. Dr. Abbott, who died in 1973 at the age of 101, is best known for his early work to develop instruments for analyzing the solar spectrum.

Early work to develop solar-powered engines was begun in the latter part of the nineteenth century by Augustin Mouchot in France and John Ericsson in the United States. Mouchot made a notable advance in collector design by introducing a truncated cone reflector that focused light uniformly along its symmetry axis, rather than at a small spot as had previously been done. This device is illustrated in Figure 4-4. In collaboration with Abel Pifre, Mouchot developed a solar steam engine that operated a printing press in Paris in 1882. John Ericsson, best known for designing the Union Navy's ironclad warship "Monitor" during the Civil War, also devoted considerable effort to solar-powered engines, building nine such engines between 1860 and 1883. He invented the Ericsson-cycle, hot-air engine for the conversion of solar heat. As illustrated in Figure 4-5, he concentrated the sun's rays through the use of a parabolic reflector. His reflector was designed to track the sun across the sky in order to maintain a constant power output. His 1883 model was claimed to develop 746 watts for 9.3 m² of reflecting surface. He concluded after his last engine that solar-powered engines were 10 times more expensive than conventional designs and therefore not economically justifiable at that time. It has been reported that by ingeniously converting his solar engines to run on fossil fuels he was able to turn the entire project into a commercial success.

FIGURE 4-3   The solar cooker of W. Adams. [Reprinted with permission from *Sunworld*, No. 5 (August 1977), pp. 17–32, published by Pergamon Press, Ltd.]

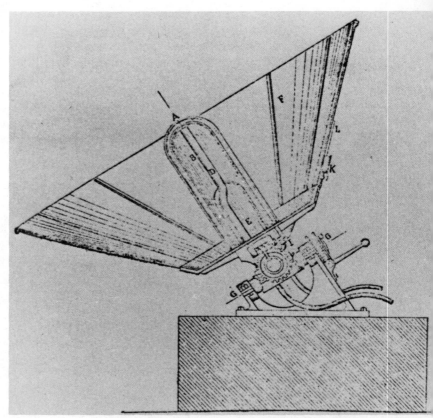

FIGURE 4-4   Mouchot's solar boiler. [Reprinted with permission from *Sunworld*, No. 5 (August 1977), pp. 17–32, published by Pergamon Press, Ltd.]

A water pump, energized by a large solar reflector, was built by A. G. Eneas in 1901 to provide water for irrigation on a Pasadena, California ostrich farm. A picture of the Eneas solar pump is shown in Figure 4-6. The solar pump, rated at 4 hp, was described in the *Arizona Republican* of February 14, 1901:

The unique feature of the solar motor is that it uses the heat of the sun to produce steam. As "no fuel" is cheaper than any fuel, the savings to be effected by the device is evident. . . . . The reflector resembles somewhat a huge umbrella open and inverted at such an angle to receive the full effect of the sun's rays on 1,788 little mirrors lining its inside surface. The boiler, which is 13 feet 6 inches long is where the handle of the umbrella ought to be. The boiler is the focal point where the reflection of the sun is concentrated. If you reach a long pole up to the boiler, it immediately begins to smoke and in a few minutes is aflame. From the boiler a flexible metallic pipe runs to the engine house near at hand. The reflector is 33-½ ft. in diameter at the top and 15 ft. at the bottom.

The entire arrangement was claimed to have cost a mere $2500, and might possibly have been a commercial success had it not been subject to an unfortunate number of mechanical failures, accidents, and eventual destruction as a result of wind storms.

FIGURE 4-5   The tracking solar moter designed by John Ericcson. [Reprinted with permission from *Sunworld*, No. 5 (August 1977), pp. 17–32, published by Pergamon Press, Ltd.]

FIGURE 4-6   Eneas' solar boiler. [Reprinted with permission from *Sunworld*, No. 5 (August 1977), pp. 17–32, published by Pergamon Press, Ltd.]

In 1908, H. E. Willsie and John Boyle, Jr. built a 15 kW slide valve engine that ran a water pump, compressor, and two circulating pumps in Needles, California. This design, shown in Figure 4-7, was based upon their earlier work using a cone-shaped reflector in Olney, Illinois. They used 93 m² of solar collector to heat water, which in turn was used to heat a volatile fluid—in this case sulfur dioxide—which was then vaporized. This device was the first to demonstrate the use of two fluids for solar power generation. Unfortunately, the cost of construction per horsepower was many times that of a conventional steam engine so that the project was a financial failure.

One man, Frank Shuman, came close to making a commercial success of a solar-powered engine. Using a flat-plate collector design of 112 m² in area he constructed a successful 3½ hp engine. The solar energy was absorbed by ether, which boiled to provide steam for the engine. Since the flat-plate collector did not have to track the sun, it was a simpler and less costly engine than other contemporary designs. Encouraged by this success, Shuman formed the Sun Power Company and convinced English financiers to back his efforts to build even larger plants. His most noteworthy effort was the construction of a solar power plant near Cairo, Egypt, aided by the technical suggestions of British physicist C. V. Boys. The latter suggested the use of parabolic reflectors placed sufficiently apart so that they wouldn't shadow each other. The solar energy was collected by black painted tubes located along the focal point of the trough-like parabolas, as indicated in Figure 4-8. Because of the concentrating reflector, the collectors were tracked across the sky to follow the sun. This was accomplished through the use of small motors directed by heat sensors and thermostatic controls. The total surface of the collectors was 1200 m², and there is authenticated information showing that the output never fell below 39 kW, and averaged around 45 kW. The system fell into disrepair during the First World War, but it has been estimated that had it been maintained properly,

FIGURE 4-7   The solar-powered plant of Willsie and Boyle. [Reprinted with permission from *Sunworld*, No. 5 (August 1977), pp. 17–32, published by Pergamon Press, Ltd.]

the irrigational services that it performed would have paid for the cost of construction within three years, after which time the only expense would have been the cost of maintenance.

One of the most practical applications of solar energy involves the production of fresh water. A 4700 m² solar still was built in the desert near Salinas, Chile, by J. Harding and C. Wilson. It delivered 23,000 liters of fresh water per clear day and operated for about 40 years. More recently, during World War II, solar survival kits were manufactured for emergency use in the South Pacific. The principle of operation of a solar still is demonstrated with the aid of the earth-water design shown in Figure 4-9. The container for the fresh water is placed at the center of the cone-shaped hole. A sheet of clear plastic is stretched over the hole with the edges held down by rocks and earth. A rock is then placed in the middle of the plastic sheet, causing it to slope downward from the sides to the center. The heat generated under the plastic causes the

FIGURE 4-8 The solar boiler of Shuman and Boys. This device developed about 45 kW and was used for irrigation purposes near Cairo, Egypt. [Reprinted with permission from *Sunworld,* No. 5 (August 1977), pp. 17–32, published by Pergamon Press. Ltd.]

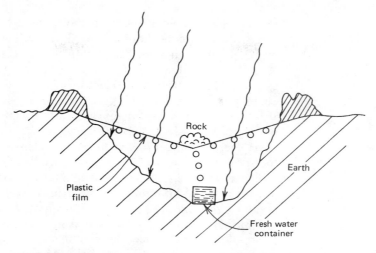

Rock

Earth

Plastic film

Fresh water container

FIGURE 4-9   The energy from the incident solar radiation warms up the space enclosed by the plastic film. This causes water in the ground to evaporate and condense on the thin plastic sheet. (Courtesy of J. I. Yellot, "Solar Radiation and its uses on Earth," published in *Energy Primer*, pp. 4–24, Portola Institute, Menlo Park, Calif.)

moisture contained in the ground to evaporate, which then condenses on the plastic sheet and runs down into the collector.

Until 1970, the only really successful commercial application of solar energy involved the installation of solar water heaters in Florida and California. Tens of thousands of these heaters were sold in both states until the middle 1950s when the advent of cheap fossil fuel energy caused their sale to decline. It was estimated that more than 50,000 such heaters existed in the city of Miami alone in 1951; but by 1970 the solar water heater had almost vanished in the United States, although it still flourished in Australia, Japan, and Israel. Recent events have now reversed this trend in this country, particularly for swimming pool applications.

In the 1930s interest spread to the topic of house heating. The pioneering experiments at the Zurich Institute of Technology were published by M. Hottinger in 1935. The incident solar radiation was captured by collectors mounted on south-facing roofs or walls with water or air as the heat transfer medium. Insulated tanks of water, rocks, or heat of fusion salts were used for energy storage. Between 1939 and 1960, Professor H. C. Hottel and his colleagues at the Massachusetts Institute of Technology made pioneering studies with a series of four different solar houses. Several other solar houses were built and tested in the period between 1948 and 1960. The most notable were the Boulder, Colorado, solar home of G. O. G. Löf who used air as the heat-exchange fluid, and the Washington, D. C. solar houses of Harry Thomason who invented the trickle-type collector. A detailed discussion of space heating, flat-plate collectors, and solar houses is given in later chapters.

As interest in solar energy applications grew, a number of international conferences and symposia were held. In 1953 a symposium on solar energy utilization was held at the University of Wisconsin under the sponsorship of the National Science Foundation. UNESCO and the Indian government sponsored a symposium on solar energy and windpower in 1954. This meeting, held in New Dehli, stressed social and political implications as well as solar applications. The International Solar Energy Society was formed in 1954 and in 1955 organized two international conferences. The first meeting held in Tucson, Arizona, stressed basic research in the field. This was immediately followed by a symposium held in Phoenix, Arizona. Over 900 people attended the conference and symposium, with 130 coming from outside the United States. In 1961, the Social and Economic Division of the United Nations organized an international conference on solar, wind, and geothermal energy applications. Over 500 scientists attended this conference, which was held in Rome.

These conferences, along with several others held during this period of time, helped to spur interest in solar energy applications. However, the time for solar energy to compete economically with conventional energy sources had not yet arrived. During the meetings in Tucson and Phoenix one prominent official predicted that by 1970 several million homes would be heated using solar energy collection schemes. Unfortunately, these glowing predictions did not materialize—the actual number was more of the order of 10.

In 1950, Chapin, Fuller, and Pearson of the Bell Telephone Laboratories developed the first solar cell for the direct conversion of the incident solar radiation into electricity. With the advent of the space age, silicon solar cells rapidly came into prominence as a means to provide power. In 1959, the first successful Vanguard satellite carried 108 solar chips, providing 0.5 W of power. By 1969, 3 million chips providing 105 kW of power were in use in space applications. The possibilities presented by further research and development in the area of photovoltaic devices are so exciting and potentially promising that a complete chapter (Chapter XI) is devoted to this subject.

B. RECENT DEVELOPMENTS

The rapid growth of interest in solar energy applications since 1970 may be partly attributed to concern over dwindling supplies of fossil fuels, but is also due in part to the growing concern over environmental problems. This escalation of interest in solar energy has not been limited just to technical and commercial development, but is also shown by the growing numbers of books, periodicals, and articles being written each year. The International Solar Energy Society, a small group for many years, has suddenly found itself deluged with membership applications.

This growth of interest in solar energy is strikingly illustrated by the increase in federal expenditures for solar energy development since 1970. During that year, essentially no federal money was being expended for research and development in the solar energy area. By 1973 the federal investment had risen to about $6 million. Under the auspices of the newly created Department of Energy, federal expenditures for 1978 are expected to reach well over 200 million dollars. The largest fraction of this will be directed toward efforts to produce solar electricity from large-scale power plants.

The generation of electricity from solar energy presents a difficult engineering challenge. With the aid of increasing federal expenditures in this field, efforts are now underway in several different areas. A detailed description of the recent developments to develop electricity using solar thermal devices is given in Chapter X, while solar cell prospects are reviewed in Chapter XI.

Solar house development is now experiencing rapid, exciting growth in all portions of the country. Development of the solar space and water heating industry has been aided by a five-year, 50 million dollar housing demonstration program. In 1976, 65,000 m² of collector area was installed! Federal estimates are that this growth rate will continue for several years, with a doubling period of less than one year. Let us look at a few examples of this solar activity. Figure 4-10 shows a solar home constructed in Tuscon, Arizona. Figure 4-11 shows a type of solar house built for use in Oregon.

Space heating with solar energy is not restricted, of course, to family residences. Figure 4-12 shows the solar heating system installed on the Timonium Elementary School in Timonium, Maryland. The solar system consists of 465 m² of collector with 57,000 liters of water for heat storage. The solar collector array has 10 banks of collectors mounted on the central windows inclined at 45°. This system was designed to provide about 50% of the heat required during December and January, and almost the entire requirement for the remainder of the heating season.

Most solar systems also provide preheating of the hot water used in the residence. This allows the solar collector to provide useful work during the entire year. The energy needed to provide hot water can often be 30 to 50% of that required for space heating. Another important contribution that can be made by solar energy is to heat water for swimming pools. Since these are most commonly used during summer conditions of hot weather, simple, inexpensive solar collector systems can be used. In response to a growing demand for this type of solar

FIGURE 4-10   The "Decade 80 Solar House" is a prototype solar residence in Tucson, Arizona, built by the Copper Development Association, Inc. The solar energy system provides 100% of the home's space heating and a majority of its air conditioning and domestic hot water needs.

heating, a large number of private companies have entered the field. Public utilities are also beginning to recognize the important contribution that solar energy can make. For example, the Public Utilities Department of the city of Santa Clara, California, will now install solar swimming pool heating systems at a very nominal cost. In its utility concept approach, the city owns the solar hardware and charges the customer an installation fee of $200 and a monthly service fee that averages about $28 per month from April to September. The monthly rates are guaranteed by the city for a three-year period, after which they can be adjusted to reflect annual cost experience in maintaining and servicing the systems.

What impact upon the nation's energy budget will solar space and water heating have in the future? Clearly the answer to this question depends upon whether or not the present growth rate in the solar industry continues for another 10 to 15 years. If this does occur, then it is easy to show that by the year 2000, solar space and water heating could provide up to 1 to 2% of the nation's energy needs. This same conclusion was reached in 1974 by the National Science Foundation.

For solar energy to attain widespread public acceptance it must be shown to be economically competitive with other forms of energy. In estimating the cost of energy for a solar system it is necessary to take into account the fact that most of the expense is for the construction and installation of the system. This initial cost, plus the small amount needed for maintenance and operation, must then be averaged in a suitable manner over the useful lifetime of the system. We do this in the following simple way. The total solar system cost is estimated and an annual, capital, fixed charge rate is determined. This is a standard business calculation of the value of the investment dollars to an individual. It could be the interest rate that the money would be earning if it were invested elsewhere rather than used to install the solar heating system, or it could be the annual cost of a loan to finance the system installation. The system installation cost for solar space heating runs about $15 to $30 per square foot for a reasonable size system. (The collector cost is about $10 per square foot for commercial systems.) In this case the cost of the solar energy can be represented by the following

FIGURE 4-11   The James solar home in Eugene, Oregon. Air is used as the heat exchange medium in this example of an active solar system.

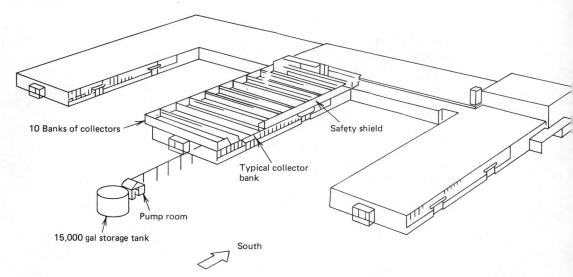

10 Banks of collectors

Safety shield

Typical collector
bank

Pump room

15,000 gal storage tank

South

FIGURE 4-12 Schematic drawing of the solar energy heating ex
periment upon Timonium Elementary School in Ma
yland. The three wings of the school are shown, wit
the collector array mounted on the central wing
(NSF-RA-N-74-126.)

simple equation:

$$\frac{\text{Solar}}{\text{cost}} = \frac{\left\{\begin{array}{l}\text{System}\\ \text{installation}\\ \text{cost}\end{array}\right\}\left\{\begin{array}{l}\text{Annual, capital}\\ \text{fixed charge}\\ \text{rate}\end{array}\right\} + \left\{\begin{array}{l}\text{Annual operating}\\ \text{and}\\ \text{maintenance cost}\end{array}\right\}}{\text{Annual energy delivered}}$$

All dollar and energy values are understood to be calculated per squar
meter (or square foot).

Let us now calculate the solar cost for the following representativ
example:

1. Annual fixed charge rate of $0.10. This corresponds to the charge fo
a 20-year loan at 8% interest.

2. System installation cost of $20 per square foot. For 800 ft^2 of sola
collector this would amount to $16,000.

3. Assume that the operating and maintenance cost is 2% per year. Thi
would be about $320 per year for an 800-ft^2 system.

4. Assume the system delivers 120 million $J/yr\text{-}ft^2$ of collector. This would amount to 96 billion J/yr for the present solar system—it is assumed that the solar system also preheats hot water the entire year.

If we put these numbers into the above equation, then the cost of energy provided by the solar system is about $20 per billion joules.

How does this cost compare with energy costs from more conventional sources? Table 4-1 summarizes the average prices for residential energy in the United States.

From Table 4-1, we see that space heating with solar energy is not competitive at present with heating by oil, electricity, or natural gas. However, this conclusion would be premature for a number of reasons. The calculation performed earlier for heating with solar energy was done by averaging over a 20-year period of time (and paying for this system over the same time period). It is certainly not fair to compare this number ($20/billion joules) with the present-day costs of other sources of energy. As can be seen from Table 4-1, very steep increases in the cost of conventional energy are expected in the next few years. A fair economic comparison would compare the above solar energy cost with some type of average cost of conventional fuels over the next 20 years. The above estimate was for an average home across the country. There is a large difference in the cost of electrical energy, which varies from about $3.80 per billion joules in the Pacific Northwest to $23 per billion joules in portions of the eastern part of the country. If one does part of the work of construction and installation, costs can be substantially reduced. In the future, mass production techniques and increased competition should bring down the costs. Also, legislation to provide relief in the form of tax credits and property tax reduction has been enacted in several states, and more progress in this direction can be expected.

Table 4-1 Average U.S. Consumer Prices for Residential Energy ($/Billion Joules)[a]

Energy Source	1970	1976	1980 Estimate
Electricity	$5.83	$8.92	$11.83
Heating oil	1.42	3.60	7.99
Natural gas	1.00	2.09	5.76

[a] Data taken from *Sunworld, 4,* 14 (May 1977).

BIBLIOGRAPHY

1. "Solar Energy Research, Staff Report of the Committee on Science and Astronautics," U.S. House of Representatives, December 1972. (U.S. Government Printing Office, Washington, D. C., No. 86-7640.)

    For the reader who wishes to quickly become acquainted with U.S. solar energy developments, this is one of the better references available. It presents the responses of the NSF, Congressional Research Service, NASA, and NBS to requests by the House Committee on Science and Astronautics for information on research presently underway in areas of solar energy.

2. F. Daniels, *Direct Use of the Sun's Energy* (New York: Ballantine, 1974).

    This is the standard reference for someone desiring an introduction to solar energy. Now available in paperback issue, this useful book contains information on a wide variety of solar topics at an elementary level. Chapter 2 contains a short review of the history of solar energy applications.

PROBLEMS

1. Consider the amount of solar energy incident upon the roof of a typical house in the Midwest in an average year. It is:
   (a) About one-tenth the average heating needs for the house.
   (b) About equal to the average heating needs for the house.
   (c) About 10 times the average heating needs for the house.
   (d) About 100 times the average heating needs for the house.
2. The Willsie-Boyle solar engine was one of the first to use the two-fluid concept. The second fluid in this engine was:
   (a) Sulfer dioxide.   (c) Methane.
   (b) Carbon dioxide.   (d) Air.
3. The early pioneering development work for the modern flat-plate collector was done by:
   (a) Archimedes.   (c) Nicholas de Saussure.
   (b) G. Buffon.   (d) Charles Abbott.
4. The failure of the commercial solar hot water heating industry in the 1950 to 1970 period can mainly be attributed to:
   (a) Inefficient flat-plate collectors.
   (b) Poor mechanical design.
   (c) Maintenance problems.
   (d) Low cost of natural gas.

5. Willsie and Boyle used what type of solar collector for their solar-powered engine.
   (a) A parabolic concentrator.
   (b) A simple flat-plate collector.
   (c) A cylindrical solar collector combined with a mirror concentrator.
   (d) A spherical receiver located at the focus of 168 small flat mirrors.

6. When a homeowner installs a small solar heating (or cooling) system upon the house, he or she is in fact creating a small-scale power station. Discuss what could be done by local and federal government to recognize this fact, allowing the homeowner to be economically competitive with other energy sources.

7. Suppose that Archimedes had no mirrors available during the Battle of Syracuse in 212 B.C. What readily available piece of equipment could be utilized as a substitute?

8. The number of new housing starts per year in the United States is about 2.5 million. Assuming that exponential growth of the solar industry will occur, with a doubling time of one year, how many years will it take before one-half of the new housing starts incorporate solar heating? About 1000 solar homes were constructed in 1976.

9. In a general way compare the cost of installing a solar hot water heating system with that of a solar space heating system. Assume that 65 ft² of collector are needed for the hot water heating system, while about 800 ft² of collector are needed for the space heating system, which also preheats the hot water. The energy needed for hot water is about one-third that needed for space heating.

10. In the example on the economics of space heating with flat-plate collectors in the text, it was assumed that the averaging process took place over 20 years. Assume that the solar system will last (with proper maintenance) for 30 years. What will this do for the economic viability of a solar heating system.

11. Discuss the proposition that one should not evaluate the cost of a solar system in the manner presented in the text, since installing a solar heating system increases the value of your house in the same manner that adding on another room increases the value of the house.

12. The present-day, average cost of energy includes both low- and high-cost contributions. Hydroelectric power is an example of low-cost energy while new coal and nuclear generating plants are examples of high-cost energy. If one installs a solar space heating system, which type of energy (from the cost standpoint) is being saved? Explain how this fact is not taken into account under present economic conditions.

13. The general public now pays for pollution abatement in the form of higher prices for automobiles and many other products. What could be done to compensate the solar energy producer for his or her role in pollution prevention.

14. Property tax relief is one way of providing assistance to the individual homeowner for installing a solar system. List at least three other ways in which the federal or state governments could provide incentives to install a solar energy system.

15. In some areas of the country, the economic cost of energy produced from conventional sources is quite low now and is predicted to increase at the rate not too much larger than inflation over the next 10 to 20 years. This places solar space heating at an economic disadvantage. The Pacific Northwest is a good example of this situation. Even so, solar hot water heating is already economically competitive in these areas. Explain in detail how solar hot water heating can be competitive when solar space heating is not.

The
V Sun

We take the sun for granted. It rises in the morning and sets in the evening without fail. In terms of humankind's residence on earth, the sun is an object that will last forever, continuously radiating away the energy that makes life on our planet possible. The aim of this book is to study in detail how the radiant energy from this, our only renewable energy source, is converted into other useful forms. Before proceeding, it is worthwhile to first answer a few questions about this very important stellar object. When and how was it formed? For how long into the future will the sun continue to radiate at its present rate? Does the sun rotate just like the earth? Does the radiant energy that we receive from the sun come from its interior or from its outermost fringes? What is the magnitude of the solar radiation that we receive? Is the magnitude of this incident solar radiation a constant over long periods of time? These are just a few examples of the many questions that can be asked about our sun.

The enormity of the universe is truly staggering. Our sun is only one of the 10^{11} stars in our galaxy, the "Milky Way." The diameter of the Milky Way is about 10^5 light-years, so that a light ray originating on one side of our galaxy takes 100,000 years to cross to the other side. Within a radius of 2×10^6 light-years are located another 15 galaxies. An example of the enormity of space is shown in Figure 5-1, which illustrates one typical cluster of galaxies. In the last 100 years or so, astronomers have searched the skies with very powerful telescopes. It is now estimated that the number of visible galaxies in the universe is

FIGURE 5-1 Photograph of the large cluster of galaxies in the constellation Hercules. A cluster may contain anywhere from two to several thousand galaxies. The average distance between galaxies is about 1 million light years. (Courtesy of Hale Observations, Catalog #68).

in the billions; there are so many that the number cannot be directly counted.

How was the universe formed? A number of hypotheses have been advanced. One of these states that the universe was created almost exactly as we now observe it. However, between 1920 and 1930, astronomical evidence that was in sharp disagreement with this model began to accumulate. It was observed that everything in the universe appears to be moving away from us. The speed of recession of an object appears to increase as the distance away from us becomes larger. Evidence for this comes from observations of the Doppler shifting of the light received from these distant objects. This phenomenon is a shifting of the color of the observed light toward the red end of the spectrum due to the motion of the source of radiation away from us. It is a very well-understood physical effect and occurs for all types of wave phenomena (sound, light, radio waves, etc.) It is as though we were at a point on the surface of an expanding balloon, with every other point receding from us as the balloon expanded. This experimental observation of an expanding universe is in sharp disagreement with the steady-state hypothesis.

About 30 years ago George Gamow proposed that the universe originated by having all its mass concentrated in a relatively small volume. The concentration of matter was so high that a temperature greater than $10^{16}°K$ was achieved. The radiation pressure from this primordial fireball was tremendous and it exploded outward. Thus, our universe was created in a big bang! This event is supposed to have occurred over 10^{10} years ago, and provides a ready explanation for the theory of an expanding universe. The parts of the fireball with the greatest relative velocities are now concentrated in the distant galaxies that we see receding so rapidly from us.

This suggestion was not widely accepted until recently when careful measurements with very sensitive detection equipment measured what is now believed to be the "background radiation" left over from the original big bang. The intensity of this background radiation varies at different wavelengths and was found to have a distribution almost precisely that to be expected if the major portion of the universe were a blackbody emitter (see Chapter 6) at a temperature of about 3.0°K. The same distribution of intensity is obtained no matter in which direction the special radio-frequency detectors are pointed. According to theory, this type of distribution is exactly what would be expected to be left over from a "big bang" occurring so long ago. Interstellar space is filled with a very low density of matter in thermal equilibrium at 3.0°K, the present remnants of the once mighty fireball.

How are stars formed from this expanding cosmic dust? The prevalent hypothesis is that occasionally some kind of shock or gravitational wave triggers the gas in a particular region of interstellar space to start forming large masses of dust. Because of the force of gravity the mass of these objects increases. Typically, this large mass of gas then condenses into a few hundred to a few thousand new stars.

The life history of a typical star is complicated and can be illustrated with the aid of the diagram shown in Figure 5-2. This is a plot of the relative brightness of a star as a function of its surface temperature.

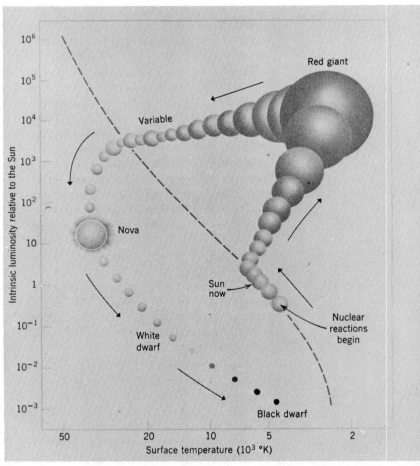

FIGURE 5-2   The life cycle of a typical star such as our sun is depicted schematically upon this Herzsprung-Russell plot. It will be another 5 billion years before the sun enters the red giant stage. (Courtesy of John Wiley & Sons, *Physics and the Physical Universe*, by J. Marion.)

Most stars are found along the main sequence, which is shown as a dashed line on the figure. The average star with a mass about equal to that of the sun spends roughly 10 billion years on the main sequence line of the diagram. At the end of this time it has used up all of its hydrogen fuel and starts to burn helium. This causes the burning rate process to increase by a factor of 10 to 100, raising the star's temperature and causing it to expand so that the surface gases are cooler. Light coming from these cooler gases has a reddish color to an observer outside the star; this is the "red giant" stage in stellar evolution. The star will increase in size and radiate away a large amount of energy with a resultant catastrophic effect upon any close lying planets. In our solar system this event will occur in about 5 billion years. After another 100 million years in the helium-burning stage, the star will completely exhaust its nuclear fuels. It will then cool off and contract because of the gravitational attraction of the cooling gases for each other. These last stages are complicated and are not as well understood. Of special interest is that occasionally the star may undergo violent eruptions with as much as 10^{-4} of the star's mass ejected into space. This is associated with a tremendous increase in luminosity. This phenomenon is called a nova, with the associated increase in brightness lasting a week or two.

The most massive stars have life histories that are quite different from those of the average stars. They shine brilliantly for a few million years and then die very spectacularly. Rather than having many relatively gentle ejections of material as do the smaller stars that undergo the nova phase, these giant stars often finish their life history very quickly in one giant explosion called a supernova. The light intensity from this event can attain a brightness 10,000 times greater than that of a typical nova. In the past 2000 years seven of these huge stellar explosions in our galaxy have been authenticated from a close scrutiny of historical records. From observations of supernovas in other galaxies, it is known that such catastrophic events are rather frequent—about one per century in a galaxy such as ours. Because this type of explosion tends to occur near the central plane of a galaxy, and since the central plane of our own galaxy is filled with obscurring dust, very few supernovas have been observed over the span of recorded history. Supernovas play an important role in the formation of the heavy elements. The heavy elements (formed by neutron-capture reactions) are ejected into space during the explosion. At some later time this heavy material is incorporated into the next generation of stars.

Our universe contains energy in many different forms such as heat, light, gravitation, and nuclear energy. Stored chemical energy, with which we are so familiar because of our large fossil fuel reserves, counts for very little in the universe as a whole. The dominant form of energy in the universe as a whole is gravitational energy. Every bit of matter attracts all other matter through the force of gravity. The gravitational energy, which separated masses possess, can be converted into light and heat by allowing the originally separated masses to fall together. Because of the large amount of mass in the universe, this form of energy greatly outweighs all other forms.

When two masses are separated the laws of physics favor their coming together under the influence of the gravitational force. If this is the favored situation it is then logical to ask why gravitational energy is still predominant after 10 billion years? Why has the universe not collapsed under gravitational contraction?

The answer to this question is difficult to give in a simple manner; essentially one can say that the universe survives because of a number of constraints (see ref. 2 in the bibliography at the end of this chapter). At the galactic level the important constraint is the large size of the universe. If a finite volume is filled with matter to a density ρ, then the matter inside the volume cannot collapse in a time shorter than the time, T, it would take a mass, m, to "fall" across the given volume under the influence of the attractive gravitational force of the rest of the mass. Using a density, ρ, of 1 atom/m³ as an average value for the entire universe, a value of $T \cong 100$ billion years is obtained. This is a factor of 10 longer than the estimated age of the universe. It appears that on a global scale the universe is safe from gravitational collapse. At the level of individual galaxies the situation is considerably different. The density inside our galaxy is a million times that of the universe as a whole, resulting in a free-fall time of only 100 million years. Since our galaxy has existed much longer than this, it is clear that some other constraint must be influencing the situation. This is the "spin" constraint, due to the orbital motion of the stars about the center of the galaxy, which is illustrated by the photograph in Figure 5-3 showing the spiral motion of stars about the center of a galaxy. In a similar way it is the orbital motion of the earth and the other planets that keeps them from falling into the sun.

**A.
SOLAR FUSION
REACTIONS**

As a typical star, the sun was formed by the mutual gravitational attraction of its constituents. When the individual particles condensed from the cosmic dust cloud and "fell" together, their kinetic energy increased, exactly as a falling stone picks up speed. As the size of the sun decreased, gravitational energy was converted to energy of motion, heating up the interior of the sun to very high temperatures, of the order of 2×10^{7}°K. What keeps the sun from collapsing even further is that at these high temperatures thermonuclear reactions involving the burning of hydrogen can take place, and these reactions release large amounts of radiation. The pressure from this radiation is sufficient to keep the sun in equilibrium at its present dimensions and to prevent

FIGURE 5-3 The constraint due to orbital motion of the stars in a galaxy is illustrated in this photograph of M104 taken with the 500-cm. telescope on Mt. Palomar. The circular paths of the individual stars about the galaxy center keep it from collapsing. (Courtesy of Hale Observatories, Catalog #32).

further collapse. About 5 billion years in the future, when the sun's hydrogen is spent and this pressure is removed, the sun will again gravitationally contract. The resulting higher temperatures will then again permit the burning of new fuels, helium and carbon, as the sun enters into its red giant stage.

What are the nuclear reactions taking place in the fiery interior of the sun? Since the sun is about 90% hydrogen you should not be surprised that the basic process involves the fusion of hydrogen to form stable helium, with the accompanying release of a large amount of energy.

This basic process, first postulated by physicist Hans Bethe, is called the proton-proton cycle. The conversion of hydrogen to helium actually involves three separate reactions. The first reaction is of particular importance to us because it consists of the interaction of two hydrogen nuclei to form deuterium (heavy hydrogen composed of a proton and a neutron). Along with the deuterium, a positive electron and a massless particle called a neutrino are also emitted. This process can be written schematically as

$$(4\text{-}1) \qquad P + P \rightarrow D + \beta^+ + \nu$$

The deuterium formed will immediately react with the hydrogen in the sun, through a nuclear interaction. The result of this process is the formation of the light isotope of helium, ^{3}He, and an energetic radiation called a gamma ray. As the ^{3}He is formed its density finally becomes large enough that occasionally two such atoms will collide and interact to form ordinary helium, ^{4}He, and two nuclei of hydrogen. These two reaction processes can be summarized as follows:

$$(4\text{-}2) \qquad D + P \rightarrow {}^3He + \gamma$$

$$(4\text{-}3) \qquad {}^3He + {}^3He \rightarrow {}^4He + 2P$$

The entire process can be summarized by saying that four hydrogen nuclei combine to form helium through the processes described by the above three reactions. This fusion process releases the energy needed to power the sun.

Why doesn't this process, starting with the reaction of Eq. 4-1, take place catastrophically with the entire sun exploding in one giant blast? The answer to this question lies in another constraint of nature. We have previously discussed examples of three of the four fundamental forces in nature; the gravitational, electrical, and nuclear forces. It is the fourth basic force, called the weak interaction, that is responsible for the reaction, described by Eq. 4-1, taking place. This weak interaction is responsible for any reaction that occurs where an electron or a neutrino is emitted. Because the weak interaction is only roughly 1 million-millionth as strong as the nuclear force, this process proceeds at about a factor 10^{-12} as fast as most nuclear reactions. This is so slow that the only way the reaction described by Eq. 4-1 can play a significant role is for a very large amount of hydrogen to be available—exactly the situation that prevails in the sun. Because of the slowness with which the above reaction proceeds, the formation of deuterium from proton-proton capture has never been observed in the laboratory.

The large amount of hydrogen that is necessary to have an appreciable energy release by this process precludes using this reaction for power on earth. The present attempts to make a controlled fusion reactor are based upon the fusion of the deuteron (D) with atoms of tritium (T), an isotope with the same electrical charge as hydrogen but with three times its mass. In principle, two protons could also interact through the "strong" or nuclear force to produce an isotope of helium, ^{2}He. However, this nucleus does not exist in nature, which leaves the reaction described by Eq. 4-1 as the only feasible way to fuse hydrogen together.

**B.
PHYSICAL
DESCRIPTION OF
THE SUN**

The sun is 1.39 million km in diameter and is at an average distance of 150 million km from the earth. The sun is a peculiar hydrodynamic object, with the equator rotating about its axis in 27 days, while the polar regions rotate about this axis once every 31 days. The temperature of the hot interior is estimated to be near 2×10^7°K, which is high enough that the fusion of hydrogen to helium can take place. This hot, interior region where most of the heat generation takes place is thought to extend out to $0.23R$, where R is the radius of the sun. It is estimated that this core region where most of the sun's fusion reactions occur contains about 40% of the mass of the sun and provides over 90% of the total heat generation.

The energy generated in the interior is transferred slowly out to the surface through a succession of radiative processes in which emission, absorption, and reradiation occur. About 1 million years is required for this energy transport to take place. The radiation that finally reaches the earth comes from a narrow, cooler surface region called the photosphere. This is a region of low-density (about 10^{-4} that of air at sea level) ionized gases that is rather opaque to visible light. Outside the photosphere are found three distinct, almost transparent regions. The first region, called the reversing layer, is a shell that is several hundred miles deep containing much cooler gases. Next is found a much thicker, less dense layer about 6000 miles in depth called the chromosphere with a temperature somewhat above that of the photosphere. The last layer is the corona, a region of very low density and very high temperature, of the order of 2 million °K. The various regions of the sun are shown schematically in Figure 5-4.

The radiation from the sun, while approximating that from a blackbody (see Chapter VI), differs slightly primarily because of the structure in the surface-emitting region and the absorption in gases at the surface. For most purposes we can consider the sun as a perfect radiator, emitting at a temperature of about 6000°K ($\cong$ 11,000°F). The wavelength range of 300 to 4000×10^{-9} meter (10^{-9} meter is called a nanometer) includes most of the energy of the solar radiation. The spectral distribution of the energy emitted by the sun is of particular importance to the useful collection of solar radiation because of the dependence of the

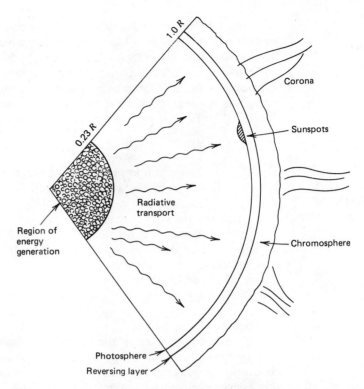

FIGURE 5-4   Schematic illustration of the various regions of the sun. As explained in the text, most of the energy generation from the fusion of hydrogen occurs in the central core of the sun.

reflective and absorptive properties of materials such as glass and various metals upon the wavelength of the radiation. The distribution of the amount of solar energy incident upon the earth as a function of wavelength is shown in Figure 5-5. About 80% of the total incident solar radiation lies in the wavelength region, which is transmitted by ordinary glass.

C. SOLAR PUZZLES

In the past 10 to 15 years numerous new developments have occurred in astronomy due to improvement in optical observational techniques and to the development of radio and infrared astronomy. Such fascinating objects as quasars, pulsars, and radio galaxies have been found. The sun has also been scrutinized in a variety of new ways. One result of all of this new, accumulated data is the indication that things within the sun are not as well understood as was previously thought. For example, the explanation of the protruding structure shown in Figures 5-6 and 5-7 is most likely that plasma is trapped in the strong, solar-magnetic and gravitational fields. Nevertheless, as they are presently understood, these fields are not sufficient to explain this unusual pattern of behavior. As new data about the sun have become available, particularly about its magnetic field, it has been conjectured that the interior of the sun may be rotating as much as six times faster than the visible surface regions. Verification of this conjecture and its implications must await further experiments.

Even the shape of the sun is now in question. As shown in Figure 5-8, the sun appears to be a perfect sphere. However, careful measurements by Dicke and Goldenburg of Princeton in 1967 indicated that the sun is slightly oblate, that is, fatter at the equator. This deviation from sphericity, if true, could have significant implications for the detailed form of general relativity. Later experiments are at variance with the Dicke result, however, so that the entire matter is an open question at present.

Perhaps the most puzzling aspect about the sun is related to our present understanding of the basic nuclear processes and hydrodynamics of the sun. Coupling these two aspects of physics together has

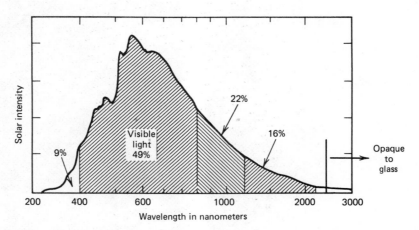

FIGURE 5-5   Division of the intensity of the solar spectrum into its different wavelength regions. About 49% of the solar energy is in the region of visible light, 9% is in the ultraviolet region, and about 4% has a wavelength long enough that it cannot pass through a thin layer of glass.

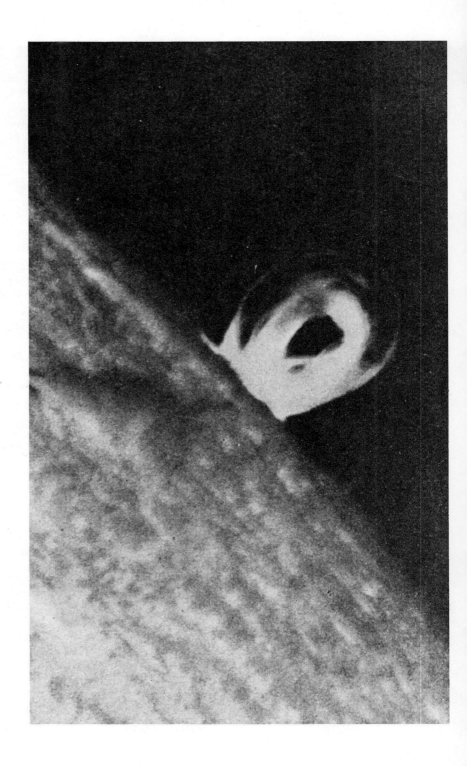

98

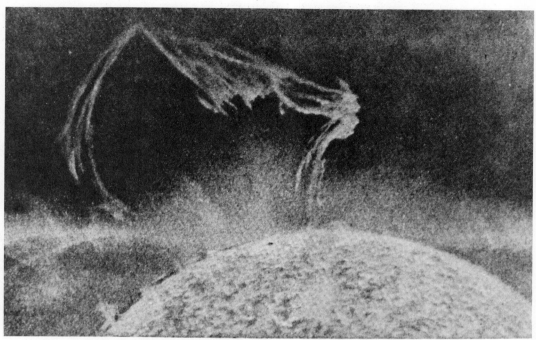

FIGURE 5-7 Puzzling solar eruption sends huge mass of helium surging up 800,000 km from the surface of the sun. Scientists are presently unable to explain this phenomenon, where a cloud of cool gas seems to run into an invisible wall in hot corona. The influence of magnetic field and gravity alone cannot account for the original explosion, stonewall effect and suspended "rain." (From: *Aviation Week and Space Technology*, January 14, 1974.)

FIGURE 5-6 Photograph of spectacular prominence taken at California Institute of Technology Big Bear Observatory. Taken from the light of hydrogen, it shows the visible surface of the sun—the photosphere—together with the lower atmosphere—the chromosphere. The pattern of markings on the surface represent a cell structure that is 32,000 km in diameter. (From: *Aviation Week and Space Technology*, January 14, 1974.)

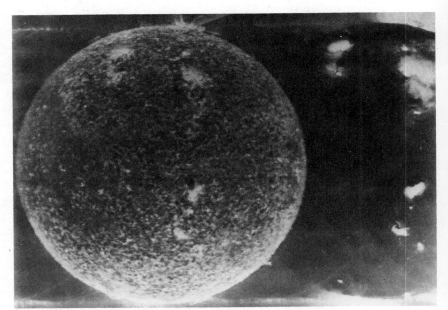

FIGURE 5-8 The solar disk appears to be a perfectly spherical object. The possibility of a slight deviation from this symmetrical shape could have important consequences for general relativity. (From: *Aviation Week and Space Technology*, January 14, 1974.)

led to an understanding of the energy generation and radiative flow inside the sun that appeared at one time to be in very satisfactory agreement with all measured properties. However, one cannot directly study the interior of the sun. The light energy generated in the core of the sun reaches the surface only after a large number of processes have occurred. For example, the light energy can collide with atoms and be scattered. Or, it may be absorbed and reemitted by the matter that makes up the bulk of the sun. These processes occur many, many times so that the light coming from the surface of the sun corresponds to light that left the center more than a million years ago. This means that when we look at the sun, the light emitted tells us only about the solar surface, and reveals nothing about the solar interior.

If the proton-proton process described earlier is the dominant fusion cycle in the sun, then occasionally an isotope of Boron, ^8B, will be formed. ^8B is not stable and will immediately undergo radioactive decay; one of the decay products is a massless particle that is called a neutrino. Unlike light, the neutrino has no appreciable interaction with matter. When a neutrino is emitted in the interior of the sun it reaches the surface 2.3 seconds later. Since it only takes 8 minutes and 20 seconds for the neutrino to cross the 150 million kilometers between the earth and the sun, it is possible to study the interior of the sun if somehow we can only detect the neutrinos. Of course, since the neutrino interacts so weakly with matter, it is very difficult to detect. Fortunately, it turns out that neutrinos have a relatively high probability (for neutrinos) of interacting with chlorine (Cl) atoms. When a neutrino interacts with a chlorine atom, an electron is expelled and an atom of argon (A) is formed. The argon atom is radioactive and easy to detect. Over a decade ago, the physicist R. Davis of Brookhaven National Laboratory initiated an experiment to measure the number of solar neutrinos striking the earth. Davis' apparatus is located 1440 m below ground level in the Homestake mine in South Dakota. His detector consists of 378,000 liters of C_2Cl_4 (ordinary cleaning fluid). Many years of painstaking measurements have led to the conclusion that the number of neutrinos detected is considerably below that expected on the basis of our present knowledge of the sun. Does this mean that we are in error concerning our present understanding of the reaction processes in the sun's interior? Or is there another explanation? A large number of hypotheses have been suggested, but again the correct answer awaits future developments.

One suggestion that has been advanced to account for the absence of solar neutrinos is that the sun is presently undergoing a period of reduced power generation. This has triggered many investigations of past historical records related to such things as sunspot activity and solar output. It is very intriguing that in 1893, E. Walter Maunder, superintendent of the Royal Greenwich Observatory in London, after a search through old books and journals, postulated that the sun was not the regular and predictable star that had previously been believed. Maunder found that between 1645 and 1715, sunspots and other solar activity had all but vanished from the sun; the total number of sunspots recorded in this 70-year time interval was less than what is seen in a single average year today. Recent studies of this time period have vindicated Maunder's findings. The time period of 1645 to 1715 was also one of practically no aurora borealis (northern lights) activity and no coronal streamers.

There is a very powerful, modern technique that allows one to shed further light upon this subject and to extend the discussion to very ancient times. Radioactive ^{14}C is formed in the upper atmosphere through the action of cosmic rays. When the sun is very active its extended magnetic fields shield the earth from cosmic rays, with the result that little ^{14}C is formed. When the sun is less active, as during the Maunder minimum, the reverse is true. It is fortunate that trees provide a record of the amount of ^{14}C in the atmosphere. By analyzing the wood in the annual growth rings of very old trees, the ratio of ^{14}C

to the common isotope of carbon, ^{12}C, can be determined. It has been found that the time period between 1645 and 1715 was a period of greatly enhanced ^{14}C intake by trees. From an analysis of the ^{14}C data over the past 5000 years it is found that there have been at least 12 solar excursions as prominent as the Maunder minimum. Another important aspect of the Maunder minimum time period is that it corresponds almost exactly with the coldest part of the "little ice age," a period of unusual cold in Europe from the sixteenth century to the early nineteenth century. During the Maunder minimum, Alpine glaciers advanced further than since the last major glacial period 15,000 years earlier. A search of old records of Alpine glacial advances and retreats correlates exactly with periods of greater or lesser sunspot activity as determined from the ^{14}C data. Therefore, it appears that for the past 5000 years, all climatological curves rise and fall in response to the long-term level of solar activity.

BIBLIOGRAPHY

1. J. B. Marion, *Physics and the Physical Universe* (New York: Wiley, 1971).

    Chapter 17 of this elementary physics text provides an introduction to the subject of astrophysics and cosmology. Material covered includes the basic stellar nuclear reactions, formation of the elements, nonoptical astronomy, stellar evolution, and the big-bang hypothesis.

2. F. J. Dyson, "Energy in the Universe," *Scientific American*, 51 (September 1971).

    This informative article on energy in the universe appears in the 1971 Energy issue of *Scientific American*.

3. N. Robinson, *Solar Radiation* (Amsterdam/London/New York: Elsevier Publishing Co., 1966).

    The first chapter of this book presents a good summary of the physical properties of the sun.

PROBLEMS

1. It is now believed that the "big-bang" hypothesis about the origin of the universe is the correct one. The best evidence for this hypothesis is:
    (a) The presence of an isotropic, blackbody radiation of temperature that is 3.0°K.
    (b) The presence of great amounts of cosmic dust in our galaxy.

(c) The presence of very high-energy cosmic rays striking the earth.

(d) The sighting of quasi-stellar objects (quasars) in the last few years.

2. Most of the energy of the universe results from:
   (a) Nuclear reactions in stars.
   (b) Supernova events.
   (c) Gravity.
   (d) Nuclear fission.

3. Our galaxy (Milky Way) would ordinarily collapse under the force of gravity if it were not for:
   (a) Radiation pressure from the center.
   (b) Rotational motion of all the stars about the galactic center.
   (c) Centrifugal force exerted by nearby galaxies.
   (d) Some unknown force.

4. The sun exists in a stable equilibrium between the force of gravitation and:
   (a) The centrifugal force due to its rotation.
   (b) The outward force due to nuclear fission.
   (c) The outward radiation pressure from fusion reactions.
   (d) The weak interaction constraints.

5. One of the following is not a useful fusion reaction for an earth-based fusion reactor:
   (a) $D + D \rightarrow {}^3He + n$
   (b) $D + D \rightarrow T + P$
   (c) $P + P \rightarrow D + \beta^+ + \nu$
   (d) $D + T \rightarrow {}^4He + n$

6. The two major constituents of the sun are hydrogen and helium. The percentage of helium in the sun is about:
   (a) 80%.   (c) 10%.
   (b) 90%.   (d) Unknown at present.

7. The "red shift" of light received from distant stars is:
   (a) Due to their gravitational collapse.
   (b) Evidence that the universe is expanding.
   (c) Early evidence that was thought to be proof of a "steady-state" universe.
   (d) Due to cosmic background radiation.

8. The direct radiation that we receive from the sun has its origin in the:
   (a) Photosphere.   (c) Chromosphere.
   (b) Corona.   (d) Sun's interior.

9. When the sun enters the "red giant" stage, the following will be true:
   (a) The sun's interior temperature will go to $5 \times 10^{6\circ}K$.
   (b) Almost all of the hydrogen will have been consumed.

(c) The radius of the sun will be greater than the orbital distance to the planet Neptune.

(d) The sun's major constituent will be carbon.

10. Most of the sun's energy results from:
   (a) Nuclear reactions.   (c) Chemical reactions.
   (b) Gravity.   (d) Fission processes.

11. In the red-giant stage a star dominantly burns which of the following?
   (a) Hydrogen.   (c) Carbon.
   (b) Helium.   (d) Silicon.

12. The energy spectrum from the sun resembles closely that of a perfect emitter radiating at a temperature of:
   (a) 5900°C   (c) 59,000°C
   (b) 11,000°C   (d) 590°C

13. The sun is located out near the rim of the "Milky Way." Why does it not fall in toward the center of our galaxy?
   (a) Because of its orbital velocity about the galactic center.
   (b) Because the force of the gravitational attraction with the rest of the stars in the galaxy is too weak for collapse.
   (c) The radiant pressure due to radiation from the center of the galaxy is larger than the attractive gravitational force.
   (d) Because not enough time has passed for this to occur.

14. If we could observe light from a star located on the opposite side of the "Milky Way," how long ago would this light have been emitted from the star?
   (a) 100 years   (c) 10 million years
   (b) 100,000 years   (d) 4.6 billion years

15. In this problem you are to match up the basic forces on the left with the examples on the right. Put the letter of the example you choose in front of the force.

	Forces		Examples
____1.	Nuclear	(a)	A falling weight
____2.	Electromagnetic	(b)	Emission of light
____3.	Weak	(c)	$P + P \rightarrow D + \beta^+ + \nu$
____4.	Gravitational	(d)	$D + T \rightarrow {}^4He + n$

16. Summarize how hydrogen is converted to helium in the interior of the sun.

17. All attempts to explain the inability to detect solar neutrinos have so far been unsuccessful. Nevertheless, a large number of reasonable hypotheses have been put forth. Although quantitative calculations must be left to experts in the field, one can speculate qualitatively about possible solutions. Give at least three plausible resolutions to the neutrino problem.

18. An interesting development in modern astronomy has been the discovery of "quasars." These bright, starlike objects have such large red shifts that astronomers generally believe they are located at the very extreme outer limits of the visible universe. At these large distances, their light output would have to exceed our entire galaxy of 10^{11} stars. Suggest a possible alternative not requiring such a massive energy usage to account for the observed light intensity.

19. An article published in June, 1976, in *Scientific American* discusses the seven possible supernova events in our galaxy over the past 2000 years. Read this article and tabulate the evidence in favor of this classification for each possible event. Would any of these criteria apply to supernova events outside of our galaxy?

20. Suppose that the interior of the sun out to $0.6R$ is rotating roughly six times as fast as the sun's surface. In this case how much faster is a particle moving at a distance of $0.5R$ than one at the surface?

21. If the "big-bang" hypothesis is true, explain how this would account for the observed fact that the most distant objects are moving away from us at the highest speeds.

22. Explain why hydrogen burning does not represent a useful fusion process for an earth-based energy source.

23. The sun and the planets of our solar system are thought to have all originated from the condensation of galactic dust about 4.6 billion years ago. Can you think of any reason why the sun is composed mainly of hydrogen, whereas the earth is now composed mainly of heavier elements, such as iron?

VI Solar Radiation

The next five chapters describe how solar radiation is collected and utilized in an efficient and practical manner. In order to understand how this important task is achieved, it is necessary to first understand the basic physical principles governing the generation and propagation of solar radiation. Since solar radiation is one example of a general phenomenon known as wave motion, this chapter begins with a discussion of waves. This is followed by a brief discussion of some of the important properties of light. The light emitted by the sun follows closely the emission of radiation by a "perfect" emitter at a temperature of 5900°K. The radiation emitted by a perfect emitter is known as blackbody radiation. To understand the operation of a flat-plate solar collector it is necessary to learn the important aspects of this type of radiation. The remainder of this chapter is devoted to studying the important practical features of radiation emitted by the sun.

A. WAVE MOTION

Consider what happens when a pebble is dropped into a quiet pond. A circular wave pattern spreads out from the point of impact. An important property of this wave is that the water does not move forward along the direction of the wave motion. A piece of wood placed upon the water moves up and down as the wave passes as shown in Figure 6-1, and does not travel along with the wave. This type of wave motion is called a transverse wave because the medium through which the wave passes vibrates in a direction perpendicular (transverse) to the direction

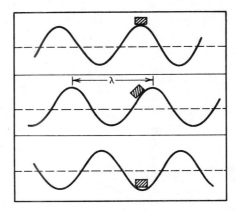

FIGURE 6-1   The movement of the block of wood illustrates that the motion of the water is transverse (perpendicular) to the direction of motion of the wave. The distance between adjacent crests (or troughs) is the wavelength, λ.

of motion of the wave. It is an important property of wave motion that the medium does not move along with the wave. Instead, it is the disturbance that we term a wave that propagates.

There is another important type of wave motion. This is longitudinal wave motion, the common example being sound. A vibrating diaphragm such as a loudspeaker, produces a wave disturbance that travels outward. While this wave motion moves outward, the air molecules move only in a localized region of space. In this case the air molecules vibrate back and forth along the direction of the wave; for this reason the wave motion is called a longitudinal wave. The similarity between the two types of wave motion just described is that in each case a disturbance travels through some kind of medium, be it air, water, or something else. Light is a transverse wave just like the water wave described above. The unique feature of light waves that distinguishes them from all other types of wave motion is that no material medium is necessary for the propagation of light.

There exists an important, fundamental relation between the wavelength of a wave and its velocity. Consider a wave source that is vibrating periodically with a time interval, T, for one complete cycle. In the case of water waves the wave generator might consist of a ruler that is dipped repetitively into the water, causing equally spaced crests to move away from the wave source with a velocity V. The speed of propagation of the waves is given as

(6-1) $V = \lambda/T$

Since the frequency, f, of periodic motion is related to the period, T, by $f = 1/T$, this relationship between velocity and wavelength can also be written as

(6-2)
$$V = f\lambda$$

This general equation is true for any simple, periodic wave; given any two of the three quantities V, f, and λ, the third is determined by this relationship.

Another useful property of waves can be obtained by observing what happens when they are reflected from some object. Consider the straight wave incident upon a diagonal, reflecting surface as shown in Figure 6-2. The angle that the incident wave makes with the reflecting barrier is labeled as θ_i; while that between the reflected wave and the barrier is shown as θ_r. Measurement of the angles θ_i and θ_r always gives that $\theta_i = \theta_r$. This relationship expressing the equality of the angles of incidence and reflection is known as the law of reflection.

B.
LIGHT

The type of wave motion of most interest to us is that of light. The sun, the stars, and electric lamps give off light. Other objects, such as the top of a desk, a refrigerator door, and this page do not. Whether a body emits light depends upon its condition. In the next section we will find

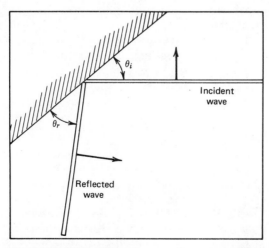

FIGURE 6-2 This diagram shows a straight wave moving toward a diagonal barrier from which it is reflected. The angle of incidence θ_i equals the angle of reflection θ_r. The direction of motion of the two rays is indicated by arrows.

that this condition of a material object usually depends upon its temperature. (It is also possible to make a light source by passing an electric current through a suitable gas. This type of light generation is due to atomic processes, which we will not examine in this text.)

Light differs from other forms of wave motion in that no physical medium is necessary for the light to propagate. It was the vibration of the particles of the medium that generated the motion of sound and water waves. In the case of light the theory of electricity and magnetism shows that light is just one form of a general type of wave motion called electromagnetic waves. Electromagnetic waves are produced whenever electric charges are accelerated. In free space (a vacuum) the velocity of all such waves is the same regardless of their frequency or wavelength. Following historical precedent it is customary to call electromagnetic waves of different frequency (or wavelength) by different names according to their origins. Figure 6-3 illustrates the different regions of the electromagnetic spectrum.

All bodies reflect some of the light that falls upon them. In the case of a thin piece of glass only a small amount is reflected, since most of the light passes through the glass. This useful property of glass is utilized in the design of solar collector systems. An object, such as a thin sheet of glass or plastic, which passes most of the light incident upon it, is said to be transparent. By contrast, light does not pass through a thin sheet of metal or wood. Materials with this property are termed opaque. Highly polished sheets of silver or aluminum will reflect most of the light incident upon them. Reflecting mirrors are used in many different optical and solar energy applications to change the direction and intensity of the light.

When light passes into or out of a transparent material it often changes direction. This bending of the beam of light is termed refraction. The

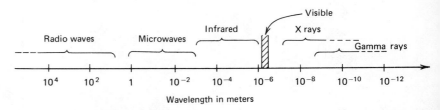

FIGURE 6-3   The electromagnetic spectrum is a continuous range of waves ranging from radiowaves to gamma rays. The descriptive names for sections of the spectrum are historical; they merely give a convenient classification according to the source of the radiation. The physical nature of the waves is the same throughout the entire frequency (wavelength) domain.

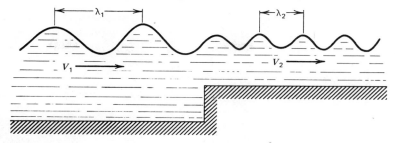

FIGURE 6-4   When water waves pass from a region of deep water to one of shallow water, their velocity of propagation is changed. The same phenomenon occurs when light passes from one medium to another. For a wave of a given frequency Eq. 6-2 shows us that the wavelength also changes.

physical basis for this bending is due to the fact that the velocity of light depends upon the type of medium. For a wave of constant frequency this means that the wavelength will be changed as the velocity changes according to Eq. 6-2. This change in wavelength is illustrated in Figure 6-4 by water waves passing from a medium defined by one velocity (in the water wave case, the velocity is determined by the wavelength and the depth of the water) to another where the wave velocity is different. Because of this change in velocity at the boundary, the direction of motion of the wave front will be different in the two media. This phenomenon is illustrated in Figure 6-5, which shows light being bent away from the normal upon passing from water to air. This phenomenon is called refraction. Analysis of this situation shows that the incident and refracted waves are related by the relation:

(6-3)
$$\frac{\sin \theta_i}{\sin \theta_r} = \frac{V_i}{V_r}$$

where V_i and V_r are the velocities of the wave motion in the two different media. Optical devices such as ordinary lenses take great advantage of this physical property.

C. BLACKBODY RADIATION

When a solid substance is heated it is found to emit radiation. A warm material emits radiation even though it often cannot be visually observed. As the temperature is increased visible radiation that can be detected by eye is emitted. The hot material will go from ''red hot'' to ''white hot'' as the temperature is raised. The same phenomenon occurs

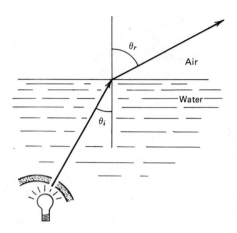

FIGURE 6-5 A wave is bent upon passing from one medium to the next where the velocity of propagation is different. In the case shown, a light wave travels more slowly in water than in air, so that a light ray in the water is bent away from the normal when it leaves.

for all objects whether solid or liquid. The energy radiated by a hot object is not monochromatic; instead it is distributed continuously over a wide range of wavelengths as indicated in Figure 6-6. As the temperature increases, more energy is associated with each wavelength region. In addition the peak of the distribution curve shifts toward shorter wavelengths. The similarity in shapes of the radiation curves for different temperatures led physicists to try and explain these phenomena with a single physical theory. Initially this was made difficult by the fact that different distribution curves were obtained at the same temperature, depending upon the nature of the emitting surface. However, it was soon found that at a given temperature, there was a maximum amount of radiation, at any specified wavelength, which could be emitted. By utilizing proper experimental techniques it was possible to obtain curves like those shown in Figure 6-6, corresponding to the maximum emission at a given temperature. When a heated object emits as efficiently as possible, regardless of the material of the emitting surface, it is called a blackbody emitter. The reason for this terminology will become apparent.

In the late 1800s, after a long series of careful measurements of the total radiation emitted by a hot, black surface, Stefan showed that the energy radiated away from the surface was proportional to T^4, where T is the absolute temperature in °K. This same relation was derived theoretically by Boltzman using very general thermodynamic reasoning. This relation between the emitted radiative energy and the temperature of an object is known as the Stefan-Boltzman radiation law and is quantitatively expressed as

(6-4) $J = \sigma T^4$

where $\sigma = 5.67 \times 10^{-12}$ J/cm²-sec-°K⁴ and J is the total energy emitted at all wavelengths.

Another important relation for solar energy collection concerns the position of the maximum in the curves of Figure 6-6. The wavelength corresponding to the maximum emission of radiant energy becomes smaller as the temperature increases. Using very general thermodyn-

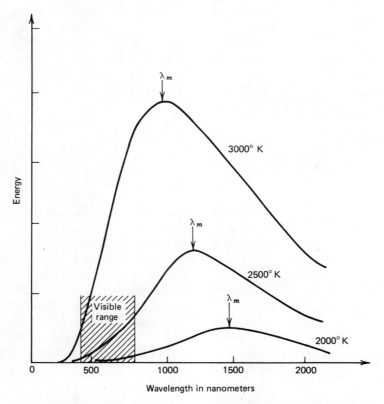

FIGURE 6-6   Distribution of energy emitted as a function of wavelength from a hot object. The solid lines show plots of the blackbody radiation curves for temperatures of 2000°K, 2500°K, and 3000°K. The wavelength at which the maximum amount of energy (per unit wavelength interval) is radiated is indicated by an arrow.

amic arguments, Wien showed that this maximum wavelength λ_m was quantitatively related to the temperature of the emitting object by

$$(6-5) \qquad \lambda_m = K/T$$

This relation, commonly called the Wien displacement law, is of fundamental importance in describing the trapping of solar energy by the "Greenhouse" effect.

Example 1

Assume that the sun is a blackbody emitter corresponding to a temperature of 6000°K. This gives rise to the solar radiation spectrum illustrated in Chapter 5 with the peak in the intensity of the radiation occurring at about 500 nm. Let us compare this wavelength with the peak wavelength when the solar radiation is emitted by the absorber plate of a solar collector, which typically operates at an average temperature of 140°F ($\cong$330°K). Using Eq. 6-5,

$$\lambda_{6000} T_{6000} = K = \lambda_{330} T_{330}$$

or

$$\lambda_{330} = 500 \text{ nm} \times \frac{6000°\text{K}}{330°\text{K}} \cong 9200 \text{ nm}$$

The blackbody distribution curve for the solar collector peaks at a wavelength far removed from the visible region, with the resulting emission consisting of radiation in the infrared region.

Our next step is to determine exactly what is meant by a black surface. To do this we consider the reflection and absorption of radiation incident normally (perpendicular) upon a thin surface as shown in Figure 6-7. Let the total radiation intensity incident upon the surface be called I_0. The amount reflected is then rI_0, the amount transmitted is called τI_0, and the amount absorbed is called αI_0. Since the total intensity incident upon the surface is either reflected, transmitted, or absorbed, we have

$$(6-6) \qquad r + \alpha + \tau = 1$$

where r, α, and τ are the coefficients of reflection, absorption, and transmission. Consider a thick surface with $\tau = 0$, so that $r + \alpha = 1$. In this case the incident radiation is either absorbed or reflected. For a given material and a specific wavelength of incident radiation, r and α have definite values. If r is known, then $\alpha = 1 - r$ is also known. If

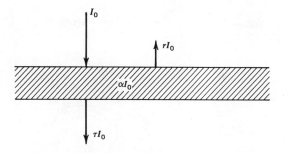

FIGURE 6-7 Schematic illustration of the reflection, absorption, and transmission of light through a thin slab of partially transparent material. The meaning of the symbols is given in the text.

we extend the discussion to the concept of radiation striking the surface obliquely, this would not change any of the conclusions to be drawn from the present discussion. It is now possible to define exactly what is meant by a white and black surface. A white surface has $r = 1(\alpha = 0)$ while a black one has $r = 0(\alpha = 1)$. That is, a black surface absorbs all of the radiation incident upon it and reflects none, while the opposite is true for a white or perfectly reflecting surface. Surfaces corresponding to real objects are never perfectly white or perfectly black. Real surfaces also often have the property that the absorptivity α is not a constant, but instead varies as the wavelength changes. This variation of α with wavelength can be utilized to design more efficient solar collectors, as shown in Chapter VII.

It is now possible to show why an ideal emitter is called a blackbody emitter. Experiments show that as α approaches unity, the emissive power of a surface approaches that given by Eq. 6-4. Let us call the relative emissive power of a given surface, defined as the ratio of its actual emissive power to that of a blackbody emitter, its emissivity, ϵ. In 1859, G. R. Kirchoff showed that at any given temperature the emitting power of a surface is directly proportional to its absorbing power. Thus, a surface that absorbs all radiation of all wavelengths would also be the best possible emitter at all wavelengths; the total radiation from it would depend solely on its temperature and not at all on its chemical or physical nature. In equation form, Kirchoff's law states that at any given wavelength,

$$(6\text{-}7) \qquad\qquad \epsilon = \alpha$$

A black object, previously defined as one that totally absorbed all the radiation incident upon it ($\alpha = 1$), can also be defined as an object that

FIGURE 6-8   The slot fire arrangement of L. Cranberg. The smallest logs are placed in front with larger logs on top and at the back. The net effect is to create a miniature black-body emitter in the form of the cavity at front. The heat is partially trapped in the cavity resulting in a higher temperature there, and the amount of radiation to the room is increased.

emits radiation at the maximum rate possible ($\epsilon = 1$). Most hot objects are not perfect emitters. The total radiation emitted by a real object is

$$(6\text{-}8) \qquad\qquad\qquad J = \epsilon \alpha T^4$$

The radiation emitted by a hot object is reduced by the factor ϵ, which is always less than 1.

It can easily be shown that the radiation emitted from a small opening approximates that from an ideal emitter. This fact has been utilized recently to design a better fireplace fire as illustrated in Figure 6-8. The standard fireplace grate is supplemented by two metal uprights at the front corners that are fitted with adjustable arms extending into the fireplace. A large log is then placed at the rear of the grate with smaller ones at the front. Another large log is placed on the moveable arms, which are adjusted so that the upper log just contacts the one in the rear of the fireplace. This creates a cavity that opens into the room; about 30% of the heat generated there is radiated into the room. A very important bonus is that the fire is easy to light. Because this arrangement traps heat so well, damp wood can be lit with only a few sheets of newspaper placed directly in the cavity.

**D.
THE INCIDENT
SOLAR
RADIATION**

Although the earth intercepts only a small fraction of the total radiation emitted by the sun, the amount received per year is equivalent to tens of thousands times the present annual energy requirement for the world. Knowledge of the incident solar flux is clearly needed for most solar energy applications. The total solar energy of all wavelengths received per unit time by a unit area of surface oriented normally to the sun's rays at the top of the earth's atmosphere is called the solar constant. The radiation incident upon this surface is illustrated in Figure 6-9. Since by definition this surface is always oriented perpendicular to the incoming solar radiation, the amount of energy received by a given area of surface should be independent of the time of year. This is the reason for calling the energy received per unit area per unit time the solar constant. The determination of this constant has been the object of a number of experimental measurements over the years. The earliest determinations, made from high mountains, had to be corrected for the transmission through the atmosphere of the different wavelength portions of the solar spectrum. More recent measurements with high-altitude aircraft, balloons, and spacecraft have permitted direct determinations outside of most of the earth's atmosphere. The most recent (1971) value for the solar constant is 1373 W/m² (1.966 cal/cm²-min, 435.5 BTU/ft²-hr).

It is more precise to say that the solar constant as defined above is determined at the average distance between the sun and the earth. The orbit of the earth around the sun, though close to circular, is actually an ellipse. This causes a variation of the solar intensity of the order of ±3.3%. The maximum intensity upon a surface oriented perpendicular

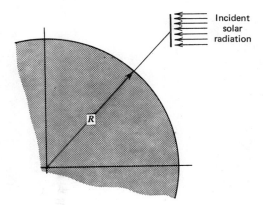

FIGURE 6-9   The solar constant is the amount of solar energy incident per unit time upon a surface of unit area oriented perpendicular to the sun's rays at the top of the earth's atmosphere.

to the sun's rays occurs in December, reaching 1.033 times the average value (the solar constant). The minimum value occurs in June when the intensity is 0.967 times the average value.

It is important to understand the reason for the changing seasons as the earth makes it's passage around the sun once a year. As shown in Figure 6-10, the seasonal variation is caused by the fact that the earth rotates around an axis that is inclined at 23.5° to the perpendicular to the plane of the earth's orbit. What is important in this case is the solar intensity upon an area parallel to the ground. In the northern hemisphere the angle between the sun's rays and the horizontal plane (e.g., at noon) is least in midwinter and largest in the summer. As can be seen from Figure 6-10, the tilting of the earth' orbit also causes a considerable variation in the duration of daylight during the year. In the northern hemisphere the maximum period of daylight is at June 22, while the minimum period of daylight occurs at December 22. The day and night periods are of equal length at the equinoxes, which come at March 21 and September 23.

From the standpoint of practical applications of solar energy one is interested in the intensity and spectral distribution at the earth's surface. In principle, one could start with the solar constant and obtain the solar radiation intensity at ground level by correcting for the attenuation in intensity due to the air mass and cloud cover of the atmosphere. This approach is seldom useful in practice because of large uncertainties in making the corrections, particularly that for the attenuation due to clouds. The most reliable way to determine the solar intensity on the ground is by direct measurement. Most commonly a measuring device

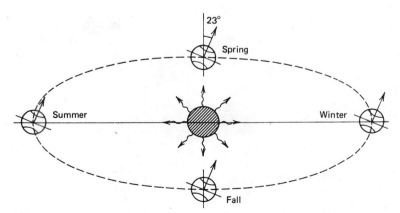

FIGURE 6-10 Graphical illustration of the reason for the changing seasons. Because of the tilt of the earth's axis, the amount of solar radiation incident upon a horizontal surface changes throughout the year.

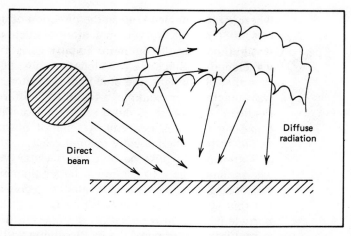

FIGURE 6-11   The incoming solar radiation is composed of two parts. The direct beam consists of radiation coming straight from the sun. The diffuse component is made up of radiation scattered one or more times by the various constituents of the earth's atmosphere.

called a pyranometer, which measures the total solar radiation (called global radiation) incident upon it, is used. While this measurement of the total solar intensity is useful it has two serious drawbacks for many specific applications. First, the solar radiation reaching the earth's surface consists of a direct beam from the sun plus a certain amount of scattered radiation (we call this the diffuse component). This breakdown is illustrated graphically in Figure 6-11. The pyranometer does not distinguish between these components. Measurements of both the direct and scattered components have not usually been available in the past so that a combination of measured data and empirical procedures is commonly used. Second, most meteorological stations orient their pyranometers parallel to the horizontal plane so that it sees the entire sky. However, for space heating and cooling applications, the desired quantity is more often the solar intensity upon a tilted surface.

The important factors affecting the observed solar intensity at ground level are the scattering by air molecules, water vapor, and dust and the atmospheric absorption by such gases as N_2, O_2, O_3 (ozone), H_2O, and CO_2. Looking first at air molecules, we note that they are very small compared to the wavelengths of radiation significant in the solar spectrum. This type of scattering of light was first investigated by Lord Rayleigh, who showed that the scattered intensity was given as

(6-9) $$I_\lambda(\theta) = I_{0\lambda}(1 + \cos \theta^2)/\lambda^4$$

where θ is the angle between the direction of incidence and the direction of scattered light. Maximum scattering occurs in the forward and backward directions. Even more important is the dependence upon the wavelength λ. Since the wavelength of blue light is considerably smaller than that of red light, the blue light is scattered much more than the red wavelengths. For example, a factor of 2 difference in wavelength would result in a factor of 16 more scattering for the short-wavelength radiation. This provides a natural explanation of the blue color of a clear sky as illustrated schematically in Figure 6-12.

Scattering by dust particles and water vapor is more difficult to assess, since their size is much larger than that of air molecules. Empirical equations have been developed to give an approximate correction for this type of scattering. While the scattering from a single, large particle is often in the forward direction, a thick cloud layer will result in multiple scattering of the incident light so that the observed intensity appears to come equally from the entire sky.

In addition to scattering, the atmosphere also absorbs some of the incident solar radiation. Absorption of short-wavelength radiation (ultraviolet rays) is due primarily to ozone (O_3). Most of this ultraviolet absorption occurs in thin layers of ozone located quite high in the atmosphere; the destruction of this ozone layer would allow the ultraviolet radiation to reach the earth's surface. There is only a small amount of absorption in the visible region. Absorption of the longer wavelength radiation (infrared) is due principally to molecules of water and carbon dioxide. The result of the various interactions with the atmospheric constituents is to reduce the intensity of solar radiation at the earth's surface to a value considerably below that of the top of the atmosphere.

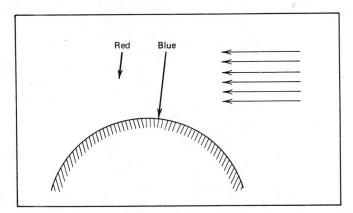

FIGURE 6-12   The enhanced scattering of light of short wavelength (blue) results in the bluish color of a clear sky.

**E.
DIRECT AND
DIFFUSE
COMPONENTS**

An attractive feature of a simple, flat-plate collector for space heating and cooling is that it will collect both the direct and diffuse components of solar radiation with almost equal efficiency. This is important for this common, low-technology application because the only region where the influence of clouds on solar radiation is negligible is in the desert. If only the diffuse component were important then the optimum orientation of a flat-plate collector would be horizontal so that it could be exposed to the entire sky. Nevertheless, a considerable fraction of the incident radiation is direct, and for this component a significant improvement in the radiant energy received can be achieved by tilting the collector toward the vertical without much loss in the collection of the diffuse component. Because of this the first step in estimating the amount of collectible solar energy is to somehow separate out the direct and diffuse components of the incident solar intensity.

Clouds are a complex phenomenon, ranging from thin, transparent cirri that attenuate the incident light only slightly to thick, thunderstorm clouds that can reduce the solar radiation to as low as 1 to 2% of its upper-atmospheric value. There are very few situations for which the type and extent of each cloud layer is observed, allowing a direct calculation of the ground level radiation. Considerable information is available, however, on the total hourly and daily radiation measured upon a horizontal surface. One can then estimate the separate percentages of diffuse and direct components in a statistically significant way by utilizing a procedure developed by Liu and Jordan of the University of Minnesota during the early 1960s. They developed an empirical procedure for estimating the percentage of diffuse and direct components, given only the average total radiation. To do this they introduced a parameter called the sunshine index, K_T. Formally, the sunshine index is defined as the ratio of the measured, total radiation upon a horizontal plane of the earth's surface to the corresponding quantity at the top of the atmosphere. That is,

$$(6\text{-}10) \qquad K_T = \Phi/\Phi_0$$

where Φ_0 is the insolation upon a horizontal surface at the top of the atmosphere, which can easily be calculated from a knowledge of the solar constant. This parameter was then empirically correlated with the diffuse component obtained from data from recording stations that also recorded both the direct and total solar radiation. From a statistical analysis of this correlation, Liu and Jordan discovered that a firm relationship existed between K_T and the diffuse component. This relationship is graphically illustrated in Figure 6-13 for the case where monthly averages were used in the analysis. The highest plotted values of K_T were near 85%, indicating that under total clear-sky conditions, about 15% of Φ_0 was lost as a result of scattering and absorption by the atmospheric air mass alone.

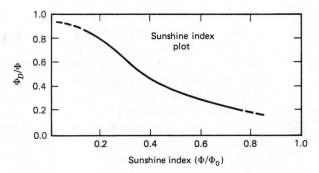

FIGURE 6-13   A graph of the empirical relation found by Liu and
Jordan between the percentage of diffuse radiation
and the sunshine index, K_T. All radiation values are
monthly averages.

At the other end of the curve where K_T is about 5%, the diffuse
percentage is about 100%. In this case almost no direct sun is seen; the
cloud cover required for this condition is very dense—perhaps four to
five miles deep. The relation outlined in Figure 6-13 is not strongly
dependent upon latitude or other factors peculiar to a particular location. If we use this graph, both the daily diffuse and direct solar components can then be estimated as follows:

1. Obtain Φ, the total energy received upon a horizontal surface, from
   the measured data obtained with a horizontal pyranometer.
2. Compute Φ_0, knowing the solar constant, the latitude, and the time
   of the year.
3. Compute $K_T = \Phi/\Phi_0$, the sunshine index that is the abscissa of
   Figure 6-13.
4. From this figure obtain Φ_D/Φ, the percentage of diffuse radiation.
5. Knowing the ratio Φ_D/Φ, one can then obtain the absolute amount
   of diffuse radiation by multiplying this ratio by Φ.
6. The direct component Φ_{DIR} is then obtained as $\Phi_{\text{DIR}} = \Phi - \Phi_D$.

In the last three steps, Φ_D is the diffuse component of the solar radiation.

**F.
RADIATION UPON
A TILTED
SURFACE**

Since a significant fraction of the incident solar radiation is direct, it is
intuitively obvious that more solar radiation will be collected by orienting the collector so that its surface is approximately perpendicular
to the sun's rays. The amount of direct radiation upon a tilted surface
can be obtained from the horizontal data using simple geometry. The

angles involved are illustrated in Figure 6-14. If the power density (J/m^2-sec) of the incident radiation is I_0, then the power density of the solar radiation upon a horizontal surface is just $I_H = I_0 \cos \theta$. If the collector surface is tilted then the intensity is $I_{DIR}(\theta_T) = I_0 \cos \theta_s$, where $\theta_s = \theta - \theta_T$. If $\theta_T = \theta$, then the intensity upon the tilted surface achieves the maximum value I_0. This minor bit of geometry illustrates the obvious fact that the maximum amount of solar radiation is received when the collector surface is oriented to be perpendicular to the incident solar radiation. The amount of solar radiation received by a tilted surface can then easily be related to that obtained for a horizontal surface:

$$(6\text{-}11) \qquad I_{DIR}(\theta_T) = I_H \frac{\cos \theta_s}{\cos \theta} = I_H \frac{\cos(\theta - \theta_T)}{\cos \theta}$$

The angle θ can be calculated from a knowledge of the latitude and the time of day. The thoughtful student will recognize that for the computation of the solar radiation received over the entire day, the angle θ will change with time, and a weighted average must be used to convert the total daily radiation measured upon a horizontal surface to the corresponding value on the tilted one.

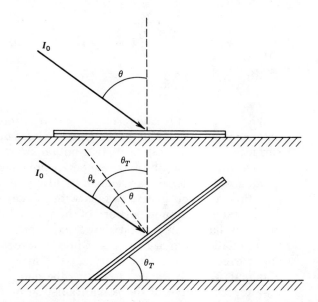

FIGURE 6-14 Showing the geometrical situation for direct solar radiation. The top portion of the figure shows direct rays incident upon a horizontal surface at an incident angle θ. When a tilted surface at angle θ_r is used, the incident radiation now makes an angle $\theta_s = \theta - \theta_r$ with this surface.

The amount of the diffuse component of solar radiation received by the solar collector also varies with tilt angle. The calculation of this fraction, assuming insolation to come equally from all points in the sky (called an isotropic distribution of solar radiation), is reasonably complicated, requiring a summation over that portion of the sky seen by the collector and including a correction for nonnormal angles of incidence. Although the calculation is tedious, the final result is simple:

$$(6\text{-}12) \qquad \frac{I_D(\theta_T)}{I_D(\text{HOR})} = \tfrac{1}{2}(1 + \cos\,\theta_T)$$

where $I_D(\theta_T)$ is the amount of diffuse radiation received by a surface tilted to the horizontal by the angle θ_T. In addition, a tilted surface also receives solar radiation reflected from the ground. For a ground area of infinite horizontal extent, whose surface reflects solar radiation diffusely, the intensity of the ground-reflected solar radiation on a tilted surface is equal to a fraction, $\tfrac{1}{2}(1 - \cos\,\theta_T)\rho$, of the intensity of the total radiation on the horizontal surface, where ρ is the reflectance of the ground for solar radiation. (ρ is often taken to be about 0.2.) Adding all three components of the solar radiation received by a tilted surface, one has

$$I_{\text{TOT}}(\theta_T) = \frac{\cos\,\theta_s}{\cos\,\theta} I_{\text{DIR}}(\text{HOR}) + \tfrac{1}{2}(1 + \cos\,\theta_T)I_D(\text{HOR})$$

$$(6\text{-}13)$$

$$+ \tfrac{1}{2}\rho(1 - \cos\,\theta_T) \times I_{\text{TOT}}(\text{HOR})$$

When we use the above techniques it is interesting to compare the pattern of solar radiation in the United States upon horizontal and tilted surfaces. The geographical distribution of radiation on a horizontal surface for the month of December is presented in Figure 6-15. Lines of equal solar intensity are shown by heavy, dark lines. We see that the radiation received generally falls with increasing latitude; the average daily total insolation in portions of the state of Washington is only about one-fifth that incident upon New Mexico or Arizona. By separating the total radiation data of Figure 6-15 into its direct and diffuse components and correcting to a tilted surface, the map shown in Figure 6-16 is obtained. The advantage of plotting the data in this manner is that it is now representative of the actual radiation upon a real solar collector. An examination of Figure 6-16 shows that the pattern for direct radiation on a tilted surface is distinctly different from that obtained for the solar radiation upon a horizontal surface. The Midwestern states are an area of high input with most of North Dakota and Montana (48° latitude) receiving more direct insolation than Washington, D.C. (39° latitude).

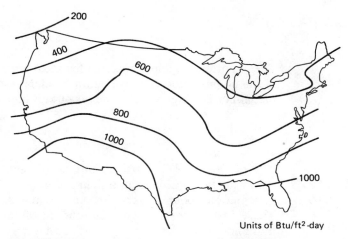

Units of Btu/ft²-day

FIGURE 6-15   Average solar radiation for the United States in the month of December as measured on a horizontal surface. The solid lines indicate regions of equal incident radiation.

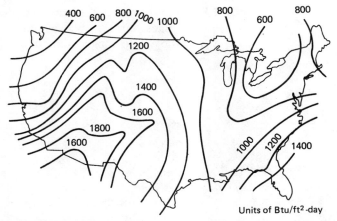

Units of Btu/ft²-day

FIGURE 6-16   Average solar radiation for the United States in the month of December as computed upon a tilted surface following the procedures outlined in the text. The solid lines indicate regions of equal incident radiation. Notice the large difference in the radiation pattern between this figure and the previous one.

125

The important conclusion is that for great portions of the central regions of the country, the solar radiation incident upon a tilted surface during the winter is roughly independent of latitude.

**G.
SOLAR
INSTRUMENTS**

The purpose of solar instruments is to measure the energy associated with radiation incident on a plane of given orientation, and to provide information about the spectral and spatial distribution of this energy. These instruments convert the energy of the incident solar radiation into another form of energy that can be measured more conveniently. It is far beyond the scope of this text to discuss in detail the many types of instruments, the problems of calibration, and the details of conversion of radiation into other energy forms. It is important only that we have some idea of the operating principles of the two most common monitoring instruments, the pyranometer and the pyrheliometer.

The pyranometer is used to determine the direct and diffuse radiation together, that is, it measures global radiation. One type of design is shown in Figure 6-17. The receiver consists of two concentric rings. The outer ring is white and the inner ring is black. The rings are often made of silver foil, 0.25-mm thick, and the coatings are magnesium oxide and Parson's black. A thermal insulator is inserted between the rings. The dull, black surface absorbs almost all the radiation incident upon it, while the white magnesium oxide reflects visible and near-

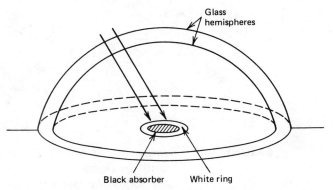

FIGURE 6-17 Pictorial representation of a pyranometer. The light absorbed by the black ring develops a temperature difference between it and the white ring concentric to it. This temperature difference is proportional to the energy absorbed.

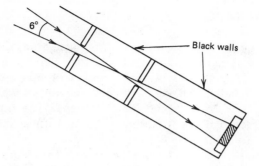

FIGURE 6-18 Illustration of the geometry used for a pyrheliometer. This instrument accepts radiation in a cone making a full angle of 6°. The sensing element at the bottom can be similar to that used in a pyranometer. The walls of the collimating tube are painted black to eliminate scattered light.

infrared radiation. The energy absorbed by the receiver generates a temperature difference between the rings, and this temperature difference is then measured by a number of platinum-gold thermocouples.

The pyranometer is used to measure the total radiation incident upon a given surface (usually horizontal). With good instruments such as the Eppley pyranometer, experiments have shown the absence of spectral selectivity, since the glass transmits practically all radiation between 350 and 2000 nm. The response is quick enough that 98% of the maximum output is reached in 20 to 30 seconds. External influences are kept to a negligible level by sealing the instruments. Pyranometers are often used for radiation measurements on inclined surfaces, in which case the calibration factor is found to depend on the position of the receiver. For the Eppley pyranometer, a change from the horizontal to vertical position may cause a 2.5 to 3% change in calibration due primarily to air convection within the bulb.

The direct beam from the sun is measured with an instrument called a pyrheliometer. This instrument is constructed with a receiving element like that for a pyranometer, and the directional sensitivity is achieved through the use of a long collimator (Figure 6-18). To reduce errors arising from imperfect alignment with the sun, pyrheliometers are constructed to include a part of the radiation from the sky immediately around the sun. The angular acceptance of a pyrheliometer is determined by the dimensions of the cylindrical tube and associated diaphragms. Typically, the instrument is designed so that the angle subtended by any point on the receiver is 5°, 41 min.

BIBLIOGRAPHY

1. *College Physics, Physical Science Study Committee* (Bedford, Md.: Raytheon Education Company, 1968).

   Chapters 3, 4, 6, 7, and 9 of this elementary text provide a detailed, but very readable account of wave phenomena and light.

2. B. Y. N. Liu and R. C. Jordan, "The Interrelationship and Characteristic Distribution of Direct, Diffuse and Total Solar Radiation," *Solar Energy 4,* 1 (1960).

   Presents relations permitting the determination on a horizontal surface of the instantaneous intensity of diffuse radiation on clear days, the long-term average hourly and daily sums of diffuse radiation, and the daily sums of diffuse radiation for various categories of days of differing degrees of cloudiness.

3. F. Y. Kondratyev, *Radiation in the Atmosphere* (New York: Academic Press, 1969).

   A thorough description of radiative processes in the earth's atmosphere, published as part of the international Geophysics Series. This book was written in English by Professor Kondratyev. Subjects covered include the measurement of solar radiation, absorption and scattering in the atmosphere, direct and diffuse components, albedo of clouds, energy distribution in the spectrum of global radiation, as well as many other topics. This book contains about 1500 references to relevant literature, although a considerable fraction of them are Russian papers.

4. M. P. Thekaekara, "Solar Energy Outside the Earth's Atmosphere," *Solar Energy, 14* 109 (1973).

   The recent availability of very high-altitude aircraft, balloons, and spacecraft has permitted direct measurements of the solar intensity outside of most of the earth's atmosphere. These measurements have been reviewed and summarized. Also included is a tabulation of the spectral distribution of the solar energy outside the earth's atmosphere, compiled from the best available data.

PROBLEMS

1. When a rock is dropped in a large pool of water, a circular wave front is observed to move out radially. The water molecules:
   (a) Vibrate horizontally along the direction of motion of the wave
   (b) Vibrate in a vertical manner perpendicular to the wave front motion.
   (c) Move along with the wave front.
   (d) Do not really move at all.

2. The velocity of sound in air is about 310 m/sec. Suppose that we "tag" an air molecule that is near the speaker of a microphone. How long will it take this molecule to reach a point 100 m distant from the speaker?
   (a) 1 sec
   (b) 10 sec
   (c) 1/10 sec
   (d) A very long time, due only to diffusion and convection of the air.

3. The wavelength of which of the following is the largest?
   (a) Light waves.   (c) X-ray waves.
   (b) Radio waves.   (d) Gamma ray waves.

4. When light waves pass from air into glass the ray describing the motion of the wave front changes direction according to Snell's law of refraction. The light ray is:
   (a) Bent toward the normal to the glass surface.
   (b) Bent away from the normal to the glass surface.
   (c) Not bent at all, that is, direction of wave front is unchanged.
   (d) Light is completely reflected at the air-glass interface.

5. Suppose that we have a source of sound waves vibrating at a rate that gives a wavelength of 10 m in air. If the vibration rate is then reduced to one-half of the original rate, the new wavelength is:
   (a) Double the original value.
   (b) The same.
   (c) One-half of the original value.
   (d) Four times the original value.

6. A transverse wave is one for which the medium vibrates:
   (a) In a direction that makes a 45° angle with the direction of motion of the wave.
   (b) In a manner resulting in complete polarization of the wave.
   (c) In a direction parallel to the direction of the motion of the wave.
   (d) In a direction perpendicular to the direction of the motion of the wave.

7. Suppose we have a surface that reflects 95% of the far-infrared radiation incident upon it. Its emissivity for this radiation is:
   (a) 5%   (c) 90%
   (b) 20%   (d) 95%

8. A pane of glass reflects 6% of the incident light and absorbs 12%. The amount of light transmitted is:
   (a) 94%   (c) 82%
   (b) 88%   (d) 18%

9. As the temperature of a hot, glowing object increases, the wavelength, λ_m, at the maximum of the blackbody radiation curve,
   (a) Decreases.

(b) Increases.
(c) Remains the same.
(d) Makes a quantum jump.
10. A blackbody absorbs all of the incident radiation. It also:
(a) Reflects radiation with 50% efficiency.
(b) Emits radiation with 0% efficiency.
(c) Reflects radiation with 100% efficiency.
(d) Emits radiation with 100% efficiency.
11. The emissivity of an object is found to be 0.87 for radiation of a given wavelength. What is its reflectivity?
(a) 1/0.87   (c) 0.13
(b) 0.87   (d) 1/0.13
12. When visible light is scattered by air molecules, which wavelengths are scattered the least?
(a) Red.   (c) Blue.
(b) Yellow.   (d) Violet.
13. Which of the following constituents of the earth's atmosphere is mainly responsible for absorbing the ultraviolet radiation?
(a) O_3 (c) N_2
(b) O_2 (d) Water vapor.
14. Compare the wavelength λ_1, at the point of maximum intensity of a blackbody spectrum from a system at a temperature of 6000°K, with the corresponding value, λ_2, for a system at a temperature of 3000°K.
(a) $\lambda_2 = 2\lambda_1$ (c) $\lambda_1 = 2\lambda_2$
(b) $\lambda_2 = \lambda_1$ (d) Insufficient information.

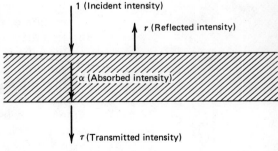

For the following three questions refer to the figure at the right. In general, $r + \alpha + \tau = 1$. (This is only meant to be a schematic illustration of the actual situation.)

15. Suppose that $\tau = 0$ (for a very thick piece of material). If the material is perfectly black,
(a) $r = 0$ (c) $\alpha + r = 0$
(b) $\alpha = 0$ (d) $2\alpha + r = 1$
16. Suppose that $\tau = 0$ as in the above problem. If the material is perfectly white,
(a) $r = 0$ (c) $\alpha = 0$
(b) $\alpha = 1$ (d) $\alpha + r = 0$

17. Suppose that we have an ideal piece of glass with $\alpha = 0$. If $r = 0.08$,

   (a) $\tau = 0.08$ (c) $\tau = 1$

   (b) $r + \tau = 1$ (d) $r + \tau = 0$

18. Which of the following is scattered the most by air molecules. Waves with a wavelength of (one micron $= 10^{-6}$m):

   (a) 0.5 micron   (c) 0.7 micron

   (b) 0.6 micron   (d) 0.8 micron

19. The intensity of the solar radiation upon a surface oriented perpendicular to the sun's rays (at all times) which is located at the top of the earth's atmosphere is largest for which of the following days?

   (a) June 21   (c) December 21

   (b) September 21   (d) March 21

20. The intensity of the solar radiation upon a fixed surface oriented parallel to the horizontal plane which is located at the top of the earth's atmosphere is largest for which of the following days? (Fixed surface located in the northern hemisphere)

   (a) June 21   (c) December 21

   (b) September 21   (d) March 21

21. Suppose that you lived in a climatic region at 45°N latitude and that clouds obscured the sky almost all of the time during the winter heating season. Assume that, under these conditions, the solar radiation was entirely diffuse. Which of the following would be the optimum orientation for your solar collector?

   (a) 90°   (c) 30°

   (b) 60°   (d) 0°

22. Suppose that the sun is located at 30° above the horizontal plane at noon on a certain day in January. Consider a flat piece of metal with an area of 4 m² that is oriented so that it faces south. For which of the following tilt angles (measured from the horizontal) will the amount of solar radiation received by the piece of metal be the largest?

   (a) 0°   (c) 60°

   (b) 30°   (d) 90°

23. A pyranometer measures which of the following?

   (a) Only the diffuse component of the solar radiation.

   (b) Only the direct component of the solar radiation.

   (c) The total amount of solar radiation incident upon the instrument.

   (d) The solar radiation received by the instrument in a cone with a half-angle of 6°.

24. What constituent of the atmosphere is primarily responsible for absorbing part of the infrared portion of the spectrum?

25. Why is a flat-plate solar collector better than the concentrating type of collector on cloudy days?

26. A sound wave is sent down a long metal bar and eventually dies out. What has happened to its energy?

27. In the following problem you are to compare the item on the left with that on the right. Put a mark in front of the one that is largest.

An example is included to clarify the procedure.

_____ Your house	__X__ Empire State Building
_____ Wavelength of light	_____ Wavelength of radio waves
_____ Velocity of light in air	_____ Velocity of light in glass
_____ Amount of light reflected by a mirror	_____ Amount of light reflected by a pane of glass
_____ Amount of light reflected by a pane of glass for normal incidence	_____ Amount of light reflected by a pane of glass with angle of incidence equal to 60°
_____ Frequency of a radio wave	_____ Frequency of a light wave

28. When you stand in front of a mirror is the image of yourself exactly like the view that others get of you? Describe the difference, if one exists.

29. You observe a rainbow. Describe the relative positions of the sun, the rain drops in the rainbow, and yourself.

30. Suppose that you are in a boat and are rowing toward a light source that you know to be submerged 2 m below the surface of the water. Suddenly you observe the light. Immediately you stop and turn around, rowing away from the light source. Soon the light source disappears from view. Knowing only that light rays bend away from the normal when passing from water to air, explain the observed phenomenon.

VII Flat-Plate Solar Collectors

For solar energy applications that involve the heating of hot water and space heating, extremely high temperatures are not needed. A liquid (usually treated water) or air heated to 110 to 140°F will suffice. For these low-temperature applications the simple, nonconcentrating flat-plate collector can be utilized. With this type of solar energy collection there is no need for the more complicated and expensive steering mechanisms that are required when concentrating devices such as lenses or parabolic reflectors are used. Another important advantage is that heat losses, which rise with temperature, are minimized. Also, the flat-plate collector will collect solar energy from the diffuse component of the solar intensity. For these reasons, solar energy systems for space and water heating have usually utilized some type of flat-plate collector.

An exploded view of a flat-plate collector using a liquid as the heat-exchange fluid is given in Figure 7-1. Most common flat-plate collectors consist of the following key components:

1. *Cover plate:*   This usually consists of one or two layers of glass or plastic.

133

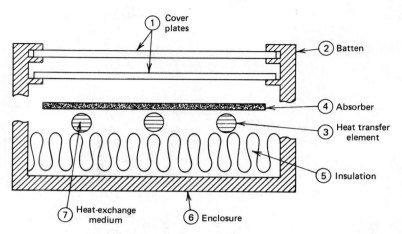

FIGURE 7-1   Exploded view of a flat-plate solar collector that utilizes tubes to transport the heat-exchange medium.

2. *Batten:* Serves to hold down the cover plate (or plates) and provide a weather-tight seal.

3. *Heat transfer element:* In the case of a liquid heat-exchange fluid, tubes are often attached to the absorber plate. This facilitates the transfer of thermal energy from the absorber plate to the heat transfer medium. Tubes, roll-band plates, open channel flow, corrugated sheets, and finned tubes are some of many different types used. The same diversity is found in air-type collectors.

4. *Absorber:* This is usually a metallic plate coated with a black paint to improve the absorption of solar radiation.

5. *Insulation:* Sufficient insulation is employed to reduce heat loss through the back of the collector to a small fraction of the total collector heat loss.

6. *Enclosure:* A container for all of the above components. The assembly is usually weatherproof. Preventing dust, wind and water

from coming in contact with the absorbed plate, tubes, and insulation is essential for optimum performance.

7. *Heat-exchange medium.*

Figure 7-2 shows schematic drawings of three other types of flat-plate
collectors. The bonded sheet design, in which the tubes are integral
with the absorber sheet, thereby guaranteeing a good thermal connec-

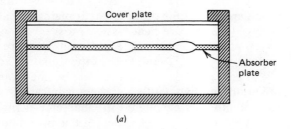

(a)

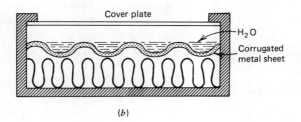

(b)

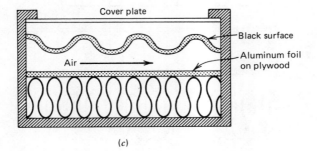

(c)

FIGURE 7-2   Examples of several styles of flat-plate solar collectors. (a) Tube-In-Sheet. (b) Thomason "trickle" collector. (c) Air collector.

tion between the plate and tubes, is shown in Figure 7-2*a*. These sheets are manufactured using copper or aluminum. A serious difficulty with aluminum is that pure water cannot be used as the heat-exchange medium because of corrosion problems; in this case a fluid containing corrosion-inhibiting agents must be used. Figure 7-2*b* shows the Thomason "trickle-type" corrugated channel collector, which is used with steeply pitched, south-facing roofs. One type of air collector is shown in Figure 7-2*c*. In this case, contact is made directly between the moving mass of air and the absorber plate. An extended surface area is used to overcome the low values of the heat transfer coefficient between the metal and air (or glass and air).

The operation of a flat-plate, solar-thermal collector can be described with the aid of Figure 7-3. Above the absorbing surface, spaced about $\frac{1}{2}$ inch apart, are one or more cover plates (usually glass) whose function is to reduce upward heat losses. Most of the sun's rays ($\cong 85\%$ for one cover plate) are transmitted through and are absorbed by the black surface. Typically, 92 to 96% of the radiation actually incident upon the blackened plate is absorbed. As the metal plate warms up it emits radiation predominantly in the far-infrared wavelength region. This wavelength was calculated in Chapter 6 using the Wien displacement law, which says that the peak of a blackbody radiation spectrum is related to the temperature by $\lambda_m T = K$. Although glass is transparent to visible light it is essentially opaque to the wavelengths larger than 3000 nanometers (nm). Thus, in a sense the sun's energy is trapped in the enclosure by the combination of a black absorber and the glazing that covers it. The cover plates warm up to a temperature between that

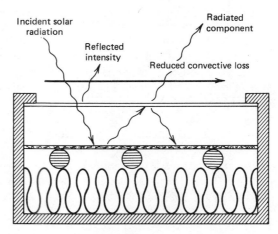

FIGURE 7-3 Schematic illustration of the greenhouse effect. The major feature is the reduced convective loss at the top cover plate.

of the hot absorber plate and that of the outside environment. This results in a substantial decrease in the radiation emitted by the collector back to the outside world. It will be shown later that while the layers of glass (or plastic) absorb and reflect some of the incoming solar intensity, the reduction in heat loss outweighs the loss in transmission, except at very low operating temperatures. Some of this heat loss reduction is due to the interception of infrared radiation by the glass, but a significant reduction also comes from the formation of blankets of relatively stagnant air. With proper spacing of the glaze, natural convection effects are greatly reduced. For example, some plastics used as covers for solar collectors are relatively transparent to the infrared radiation, but still insulate effectively because of the reduction in convective losses. The topmost cover plate greatly reduces heat losses due to direct wind convection, which is by far the dominant heat loss mechanism when compared to the situation where no cover plate is used. A further reduction in the heat loss can be obtained through the use of a selective surface that has a low emissivity for the infrared and a high absorptivity for the incident solar radiation spectrum.

Glass has been the principal material used to glaze solar collectors because of its high transmittance (about 90% for high-quality glass of low-iron content) and almost complete opacity to the infrared radiation emitted by the hot absorber plate. The transmittance of a sheet of glass varies with the angle of incidence of the light upon it, primarily because of the rapid increase of the reflected intensity at large angles. This is illustrated in Figure 7-4. For angles of incidence less than 50°, common glass reflects about 8% of the incident radiation with absorption accounting for another 4 to 6%. There are available techniques for making glass with a total transmissivity in excess of 95% for near normal angles of incidence. One possibility involves etching the surface in order to provide a gradation in density; this has the effect of reducing the reflected intensity. The same effect can also be achieved with a thin coating of a material other than glass. Unfortunately, low-reflectivity glass is rather expensive. The amount of absorption in the glass can also be reduced by lowering the iron content. Detailed calculations with complete solar systems show that the decrease in system performance is considerably less than the reduction in transmissivity; a reduction of 1% in the absorption in the glass will result in only a $\frac{1}{2}$% increase in the useful energy obtained from the system. This is because the energy absorbed in the glass is not all lost from the collector, since this absorption raises the temperature of the glass, thereby reducing the convective and radiative losses from the absorber surface. Plastic films also have a high transmittance for the solar spectrum and have often been used as cover plates. Their transmittance curve, shown in Figure 7-4 for mylar, is often better than that of glass. Some disadvantages include a reduced opacity to long-wavelength radiation, a limited ability to

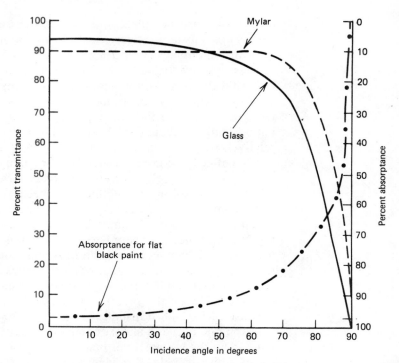

FIGURE 7-4 Transmission of light through glass and mylar as a function of the angle of incidence. The angle is measured from the normal to the glass or plastic surface. Also shown is the absorptance of flat-black paint.

withstand higher temperatures, and most types undergo changes resulting in loss of transmission due to the sun's ultraviolet radiation. The effect of dirt and dust on collector glazing appears to be small, and the cleansing effect of an occasional rain seems to be adequate to maintain the transmittance within 2 to 4% of its maximum value.

The absorptivity of the absorber plate also varies with the angle of incidence (Figure 7-4). The common choice for coating the absorber plate is a flat, black oil-based paint, with an appropriate thin undercoat of primer. The primer should be of the self-etching type to avoid the peeling off of the paint after a few years of operation. In any case, collector surface absorptance should not be less than 90%. As with the previous case concerning absorption of solar radiation in the glass, a reduction in the absorptance of the black absorber does not lead to a corresponding reduction in the system performance. It has been found that a 1% decrease in the absorptance will result in only about a $\frac{1}{2}$%

decrease in useful solar output. Again, this is due to lower heat losses in the collector since the system is running somewhat cooler.

The choice of material to use for the absorber plate is a serious one. Copper (Cu), steel, and aluminum (Al) have been by far the most common choices. No definitive answer can be given, since under proper conditions all three will work satisfactorily. Copper and some types of stainless steel have the advantage that they can be used with untreated tap water, whereas neither aluminum nor carbon steel should be used in this environment under any circumstances. In the case of copper flow, velocities must be chosen to avoid corrosion due to erosion. In Table 7-1 some of the useful characteristics of the three common materials are compared for the case of plate and tube absorber combinations. In the computation of the energy used for the steel absorber plate, the tube length was assumed to be 50% greater than that for the other two materials because of reduced tube spacing required to give equal performance (reflecting the low thermal conductivity of the steel). Steel sheet could be used in a bonded style as in Figure 7-2a. The resulting large contact area would reduce greatly the importance of the low thermal conductivity of steel. The largest fraction of the energy cost for the collector comes in the manufacturing of the absorber plate. Regard-

Table 7-1   Comparison of Absorber Plate Materials[a]

Item	Copper	Steel	Aluminum
Specific gravity (lb/in.^{3})	0.82	0.28	0.10
Thermal conductivity (Btu/hr-Ft-°F)	226	27	122
Plate thickness for equal performance (in.)	0.016	0.064	0.032
Weight per square foot for equal performance in pounds (6-in. tube spacing for Cu and Al; 4-in. spacing for steel)	1	3.9	0.6
Flow correction factor, F' (2-in. tube spacing)	0.996	0.940	0.960
Energy cost per pound (Btu)	47,400	11,600	116,690
Energy cost per square foot (Btu)	47,000	45,200	70,000

[a] Data for this table obtained from articles in the May, 1977 issue of *Solar Age* magazine. All data are presented in English units.

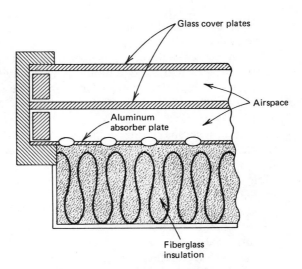

Glass cover plates

Airspace

Aluminum absorber plate

Fiberglass insulation

FIGURE 7-5 Schematic illustration of the edge details of a flat-plate collector.

less of the choice of absorber material, the energy used in the manufacture of the flat-plate collector is less than the energy that will be collected by the solar system in one year of operation.

The construction details of flat-plate collectors vary considerably; examples can be found by obtaining the flat-plate collector literature from the many commercial companies now in this business. Figure 7-5 illustrates one type of arrangement of glazing, absorber plate, tubes, and insulation for a solar collector. The optimum spacing between windows has been experimentally and analytically determined to be about 1.0 to 1.5 cm. As another example the collector arrangement used by Revere Copper is shown in Figure 7-6.

A. COLLECTOR HEAT LOSSES

There are three different physical mechanisms by which heat may be transferred from one place to another. These processes are conduction, convection, and radiation. Conduction is the transfer of heat through a quantity of material due to the molecular motion. For example, if a rod is heated at one end, the molecules there vibrate faster, with the result that energy is transferred down the rod because of successive molecular collisions. Metals are good conductors of heat, whereas materials such as glass, ceramics, wood, and gases are poor conductors of heat. Convection is the process in which a warm liquid or gas moves from one place to another, thereby effecting a transfer of energy. We have already considered radiation, which involves the transfer of energy by means of electromagnetic waves.

In a flat-plate collector the principal heat loss mechanisms are the following:

1. Conduction loss from the back of the absorber plate through the insulating material.

2. Conduction losses through the side of the collector.

3. Convection losses upward through the cover plates.

4. The upward radiation loss.

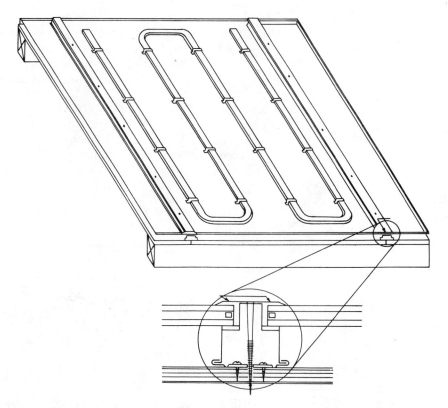

FIGURE 7-6   Flat-plate collector design of Revere Copper. Cross-sectional view shows details of insulation between the absorber plate and the glazings.

The division of the heat losses from a flat-plate collector are shown schematically in Figure 7-7. The rate of energy loss through the back insulation per unit area is directly proportional to the thermal conductivity K and the difference of temperature between the absorber plate and the environment in back of the collector frame, and is inversely proportional to the thickness, D, of the insulation. This is mathematically expressed as

(7-1)
$$Q_{Back} = \frac{K}{D}(T - T_{Back}) = \frac{K}{D}\Delta T_{Back}$$

where T is the temperature of the absorber plate and T_{Back} is the temperature at the back of the collector. Since it is easy and economical to provide good insulation on the back of the solar thermal collector, this heat loss is usually small compared to the losses by convection and radiation of the top surface. Typically, 7 to 10 cm. of fiberglass insulation is sufficient for this purpose. Equation 7-1 can be written in the following form:

(7-2)
$$Q_{Back} = U_{Back}\Delta T_{Back}$$

where U_{Back} is called a heat transfer coefficient. The heat transfer coefficient, U_{Back}, is then a measure of the heat loss through the back insulation. A well-designed collector will have U_{Back} as small as possible, consistent with the economics of the situation. For most collector designs, the heat losses through the sides and the back are small com-

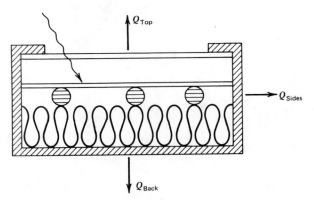

FIGURE 7-7   Schematic drawing of the heat losses in a flat-plate solar collector. The total heat loss is the sum of the losses through the sides, back, and top of the collector. In a properly designed collector the loss through the top is dominant.

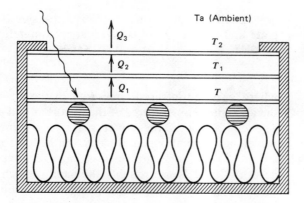

Ta (Ambient)

Q_3 T_2

Q_2 T_1

Q_1 T

FIGURE 7-8   Cutaway through a flat-plate solar collector indicating the transfer of heat from the hot absorber plate. At equilibrium each of the heat flows—Q_1, Q_2, and Q_3—are identical.

pared to the losses through the top cover plate. A useful approximation is that the heat losses through the sides and the back are about 15% of the upward heat loss.

The upward heat loss is schematically illustrated in Figure 7-8, showing the heat losses Q_1, Q_2, and Q_3 between the absorber and second cover plate, between the second and first cover plate, and between the first cover plate and the outside world, respectively. (This simple picture neglects minor corrections such as the energy absorbed from the incident solar radiation by the cover plates.) This upward heat loss is mainly due to convection and radiation. Between the black absorber plate and the adjacent cover plate, and between the two cover plates, the convection and radiation losses are roughly the same. To reduce the interior convection, some flat-plate collectors have been constructed with a honeycomb-type of cellular structure. In cylindrical glass collectors, this convection loss can be eliminated by evacuating a portion of the collector. For the cover plate that is in contact with the outside world the dominant loss mechanism is convection. At a wind speed of 9 to 15 kmph this convection loss can exceed the radiative one by more than a factor of 5. Consequently, flat-plate collectors are always constructed with at least one glazing.

It was shown in Chapter VI that the total energy radiated by a hot object is proportional to the fourth power of the absolute temperature. Two surfaces at different temperatures will then have a net amount of energy transferred between them. The calculation of this energy transfer between two parallel plates is quite straightforward but, since it involves summing up infinite series, is beyond the level of this text. The important result is that the net radiative transfer between the hot absorber plate and a cover plate is given approximately as

$$(7\text{-}3) \qquad Q_{RAD} = \epsilon_p \sigma (T^4 - T_1^4)$$

where ϵ_p is the infrared emissivity of the absorber plate and T, T_1 are the absolute temperatures of the absorber and cover plate, respectively. The emissivity of the cover plate is assumed to be 1 for the long wavelength radiation emitted by the hot absorber surface. The calculation of the convection loss is more involved and will not be given here. If T and T_1 are roughly the same, which is the case for flat-plate collectors, then $T^4 - T_1^4$ is approximately proportional to $\Delta T = T - T_1$. In this case,

$$(7\text{-}4) \qquad Q_{RAD} = U_{RAD}\Delta T$$

which is a form just like that found in Eq. 7-2. The convective loss can be approximated by a similar form so that the total amount of heat transferred from one surface to another can be written as

$$(7\text{-}5) \qquad Q = U\Delta T$$

for each pair of surfaces (considering the outside world as a surface). Finally we note that if we know the individual U's, we can directly calculate an overall heat transfer coefficient U_{TOP}:

$$(7\text{-}6) \qquad Q = U_{TOP}\Delta T_{TOP}$$

where $\Delta T_{TOP} = T - T_a$ is the difference in temperature between the absorber plate and the outside world. The calculation of U_{TOP} is one of the important tasks facing an engineer interested in designing a flat-plate collector.

The operation of a flat-plate collector can now be simply described using Figure 7-8. We assume that the collector has come to equilibrium so that all temperatures and heat flows have assumed constant values. The incoming solar radiation is absorbed by the black plate, which heats up, reaching an equilibrium temperature T (assumed uniform in our treatment), which is determined by the rate at which heat is carried away by the water and by the rate at which heat is radiated and convected to the adjacent cover plate. (Remember that the radiation emitted by the absorber plate is of very long wavelength and is almost entirely absorbed by a glass cover plate or partially absorbed by a plastic cover plate.) The cover plate then comes to an equilibrium temperature T_1 lower than T. Heat is then transferred from this cover plate to the next, which comes to an equilibrium temperature T_2 lower than T_1, but is still somewhat higher than the outside ambient temperature T_a. This

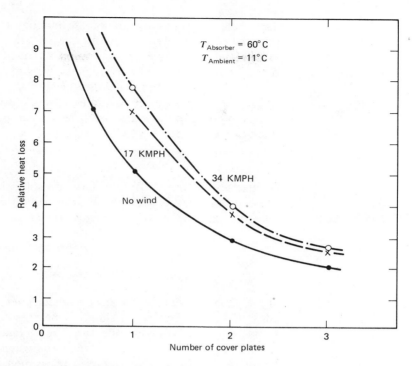

FIGURE 7-9   Showing the reduction in upward heat loss as the number of cover plates is increased.

top cover plate then loses heat to the outside world by radiation and convection, with the latter term strongly dominant when there is a wind.

It is now apparent why it is advantageous to utilize cover plates. At equilibrium the amount of heat transferred between adjacent pairs of surfaces must be the same; otherwise we would not be in equilibrium and the temperatures would change. This means that $Q_1 = Q_2 = Q_3 = Q$. The heat transferred from the cover plate to the outside world, Q_3, is the same as Q_1. The latter quantity is greatly reduced over what it would be if there were no cover plates because Q_1 is proportional to $\Delta T_3 = T_2 - T_a$, which is smaller than $\Delta T = T - T_a$. Stated in another way, the heat loss from the collector is reduced through the use of cover plates because the temperature of the surface in contact with the outside atmosphere is reduced. Besides cost, the only negative feature of cover plates is that the transmission through them is not perfect, so that a compromise must be struck between the reflection and absorption loss, which increases with the number of plates used, and the upward heat loss, which decreases as the number of cover plates increases.

The effect of increasing the number of cover plates is graphically portrayed in Figure 7-9, which shows the relative heat loss of a flat-plate collector as a function of the number of cover plates. With an assumed temperature of 60°C for the absorber plate and 11°C for the outside ambient temperature, it is seen that the convection losses due to wind striking the outer cover plate are still quite large for one cover plate. When two cover plates are used, the heat loss is strikingly reduced, and the sensitivity to wind is also greatly reduced. Only a minor improvement is obtained when three cover plates are introduced; in practice the number of cover plates is almost greater than two.

B. COLLECTOR PERFORMANCE

The useful heat output from a flat-plate solar thermal collector is equal to the difference between the heat absorbed and the heat lost by the collector and can be written quantitatively as

$$(7\text{-}7) \qquad \begin{aligned} Q_u &= \text{heat absorbed} - \text{heat lost} \\ &= \tau\alpha\,\Phi_o - U\Delta T \end{aligned}$$

where U is the sum of the heat transfer coefficients through the top, sides, and back. This result simply expresses conservation of energy as it relates to the flat-plate collector. The incident solar radiation is Φ_o, and ΔT is the temperature difference between the absorber plate and the outside ambient temperature. The product $\tau\alpha$ is the product of the resultant transmittance through the collector cover plates with the absorptance by the black absorber plate averaged over the angles of incidence of the incoming solar radiation over some period of time. With no cover plates τ is one, and for low ΔT the useful heat collected will be greater than with cover plates. However, U will be much larger in this case, particularly if there is any wind, and as ΔT increases the useful heat collected will be larger if cover plates are used.

It is instructive to rearrange Eq. 7-7 by dividing through by the incident solar flux Φ_o. The ratio of Q_u to Φ_o gives the fraction of the incident insolation that is captured by the solar collector and is termed the collector efficiency η.

$$(7\text{-}8) \qquad \eta = \frac{Q_u}{\Phi_o} = \tau\alpha - \frac{U\Delta T}{\Phi_o}$$

Typically, about half of the incident energy is absorbed and converted into useful energy; this means that the efficiency η is 0.50, or 50%. When $U\Delta T$ is greater than $\tau\alpha\,\Phi_o$, the incoming solar radiation can no longer provide for the collector losses. This will occur when the collector is warm, the outdoor air is cold, and the incoming radiation intensity is low. Thus, when $\eta = 0$, the heat losses are equal to the energy absorbed and no useful energy is obtained.

Clearly, the efficiency of a solar thermal collector is strongly affected by the difference between the temperatures of the absorber plate and the outside atmosphere. It is desirable to hold the absorber plate temperature low to obtain high efficiency, but at the same time a reasonably high temperature is needed to supply heat to a building at a practical working temperature. Figure 7-10 illustrates how the performance of a flat-plate collector goes down as the temperature increases and as the insolation, Φ_o, decreases. This demonstrates again the advantage of using two cover plates when ΔT is large. The efficiency drops off most rapidly with ΔT for the case of no cover plates. All of the incident solar energy reaches the absorber plate when no cover plates are used because of no transmission loss. However, at higher values of ΔT, the convective heat loss dominates and the addition of cover plates greatly reduces this loss. From Eq. 7-5 we can also see how to empirically obtain U. By measuring Q_u, ΔT, and Φ_o under varying conditions one can trace out the collector efficiency curve. The slope of the straight

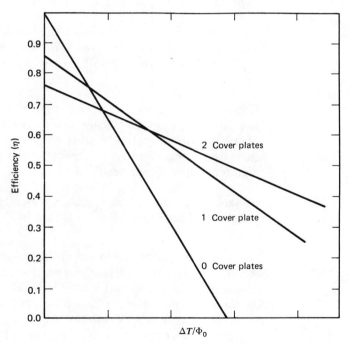

FIGURE 7-10 Plot showing how the efficiency of a solar collector varies as a function of the temperature difference between the absorber plate and the outside atmosphere as well as the intensity Φ_o of the incident solar radiation. The slope of the efficiency waves is the heat transfer coefficient, U.

line through these points then gives U, without recourse to complicated calculations.

The use of efficiency curves to compare different collectors must be viewed with caution. Some collectors have good efficiency curves when evaluated at noon, but fall off rapidly as the rays from the sun make larger angles with the normal to the collector. In fact, one is usually interested in the average performance over an entire day. Similarly, the efficiency measured during the winter months is most relevant for space and water heating applications, whereas the performance in summer is most relevant for cooling purposes.

There are many ways of improving collector performance.

1. Coat the black absorber plate with a special surface called a selective surface. This can significantly reduce the upward radiative loss in the collector. Selective surfaces are discussed in more detail in the next section. However, the cost of such surfaces is high, and their lifetime expectancy is very uncertain at present.

2. Apply an antireflection coating to the surface of the collector glazing. The reflection loss (about 16% for a double-glazed reflector) can be reduced by a factor of 2 to 5. Again, these coatings are costly and not durable.

3. Use evacuated glass collectors to reduce the conduction and convection losses. To do this requires special efforts to withstand atmospheric pressure. Presently glass collectors are quite expensive and their use appears limited to applications such as solar cooling where high temperatures are required.

4. Enhance the amount of radiation entering the collector by using reflectors. This approach is discussed in Chapter VIII.

We conclude this discussion of collector performance by noting that thus far the temperature of the absorber plate was assumed to be constant over its entire surface. This is far from true. In the type of collector design illustrated in Figure 7-8, the temperature of the absorber plate is hotter between the tubes carrying water than it is at the tubes themselves. This nonuniform distribution is essential so that there will be a flow of heat from the hot absorber plate into the cooler heat transfer fluid. It is this heat that is carried away by the fluid that is used to provide practical space heating and cooling. The temperature distributions between pipes for this style of flat-plate collector design is schematically illustrated in Figure 7-11. The temperature also varies along the flow direction as the fluid warms up as it traverses the plate.

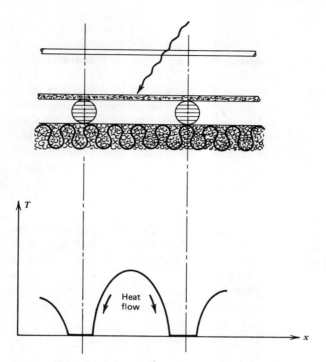

FIGURE 7-11   Cutaway drawing of a solar thermal collector. This illustrates roughly the temperature distribution between the fluid-carrying pipes. The temperature of the absorber plate is highest midway between the water tubes.

This results in a roughly linear distribution of temperature from inlet to outlet. Since the average temperature across the absorber plate for such a linear distribution is just

$$(7\text{-}9) \qquad\qquad T_{\text{av}} = \frac{T_{\text{in}} + T_{\text{out}}}{2}$$

this definition is often used in calculating $\Delta T = T_{\text{av}} - T_a$, with T_{av} computed from Eq. 7-9. As the inlet and outlet water temperatures are commonly monitored, this usage has the practical advantage of defining the collector efficiency in terms of easily measured quantities.

This nonuniform distribution of temperature across the collector causes a modification to Eq. 7-7. An extensive mathematical analysis shows that Eq. 7-7, for the net heat collected, becomes

$$(7\text{-}10) \qquad\qquad Q_u = F'(\tau\alpha\,\Phi_o - U\Delta T)$$

where F' is the ratio of the actual heat collected by the solar collector to the heat collected if the entire solar collector were at the average fluid temperature. For most reasonable collector designs of the water type, the flow correction factor F' is of the order of 0.90 to 0.95. Collector designs using air as the heat-exchange fluid often have F' factors that are considerably smaller. However, the improved heat transfer to the building being heated, obtained with air systems, appears to overcome the disadvantage of having a lower efficiency for the useful heat collected by the flat-plate collector. The flow correction factor F' is a function of a number of parameters such as the tube diameter, tube spacing, flow rate, etc.

C. SELECTIVE SURFACES

It has been noted before that one way to reduce the upward heat loss in a solar collector is to use an absorber surface that is black to the incident solar radiation, but emits very weakly in the long-wavelength region. Figure 7-12 shows the radiated intensity as a function of wavelength for both the solar spectrum and for a typical absorber plate (about 93°C or less). It is seen that there is essentially no overlap between the two distributions.

A selective surface is one that absorbs the incident solar energy, but suppresses its thermal radiation. Thus, if α is about 0.94 in the visible and near-infrared regions (300 to 1200 nm) and is less than or equal to 0.1 for wavelengths greater than 1200 nm, a large reduction in the radiative energy loss will be obtained. This is illustrated by looking at

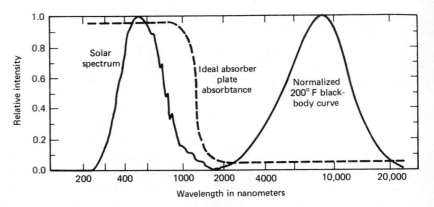

FIGURE 7-12   A diagram of the relative intensity distributions for the solar spectrum and for a typical absorber plate in a flat-plate collector. An excellent selective surface would be almost black in the solar region and have almost no absorptance in the long-wavelength region.

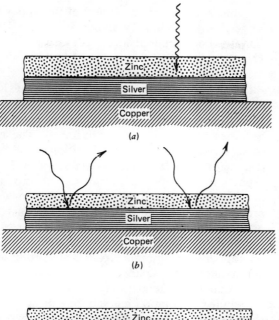

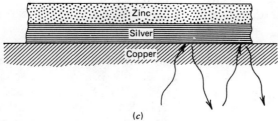

FIGURE 7-13   Illustrating the behavior of a selective surface. See
text for a discussion of the principles of operation.

the behavior of a nearly ideal selective surface as shown by the dashed
line in Fig. 7-12. The absorptance is almost unity below 1200 nm and
almost zero above this wavelength. Since $\alpha + r = 1$, another way of
looking at this is to say that the reflectance is almost zero below 1200
nm and almost one for wavelengths greater than this.

The principle of operation of a selective surface is illustrated in Figure
7-13. The copper absorber plate is covered by thin layers of zinc and
silver in this type of selective surface. The solar radiation incident upon
the surface is absorbed by the layer of zinc. Thus the absorptivity in
the visible and near-infrared wavelength regions is almost unity. However, as shown in Figure 7-13b, radiation in the infrared region (corresponding to emission from a hot absorber plate) passes through the thin
layer of zinc and is reflected by the thin layer of silver. This means that

the emissivity (or absorptivity) of the upper layers is very small for the longer wavelengths. As shown in Figure 7-13c, the infrared radiation from the hot copper absorber is mostly reflected back into the absorber. The net result of all this is a surface with $\alpha \cong 0.05$ to 0.20 in the infrared wavelength region.

One way to obtain selectivity is to use the high-infrared reflectance of a metal such as silver or aluminum to obtain low emissivity, and to overlay this with a material that strongly absorbs the solar flux but becomes transparent in the infrared. These properties are shown by some semiconductors and transition metal oxides and sulfides. Historically, following the initial suggestion of H. Tabor, a pioneer in many aspects of solar energy, absorber-reflector tandems of these materials have been the predominant configurations used in all but a few approaches to selective surfaces over the past 20 years. The most popular of these coatings are known as selective blacks and are composed of nickel-zinc-sulfide deposited on nickel or galvanized iron, copper oxide on copper or aluminum, oxides of chromium or chromium-nickel-vanadium alloys. These are all economically fabricated by electroplating or electrochemical deposition.

Detailed simulation calculations of actual solar systems have shown that significant improvement can result through the use of selective surfaces. For example, a single-glazed solar system in Bismark, North Dakota, with a selective surface (emittance of 0.10 for the far infrared) would require only 62% of the collector area needed for a single-glazed collector with no selective surface. These calculations assume that the solar system provides 75% of the total space heating load. For Medford, Oregon, similar calculations have shown that the selective surface system requires only 70% of the collector area needed without the selective surface. The corresponding percentage for sunnier areas is higher with a value of 84% for Phoenix, Arizona. In all cases, performance was actually reduced by the use of double glazing over a selective surface as compared to single glazing over a selective surface. This is because the increased absorption and reflection caused by the second layer of glass results in a greater reduction in system output than the corresponding gain resulting from decreased convective and radiative losses. In summary, if the economic costs are justifiable, it is advantageous to use a selective surface with one glass cover plate. If two cover plates are used then no selective surface should be utilized.

D. HOT WATER HEATING

In the early 1930s in the United States, widespread commercial use of solar energy for hot water heating was made in Florida and California. The collectors usually consisted of a single glass cover plate mounted over a metal box containing blackened copper tubes. Tens of thousands of these heaters were sold in both states until about 1960. With the onset of cheap natural gas, the solar water heating industry in the United States declined rapidly after this. By 1973, the industry had almost vanished; only to see a sudden resurgence of interest as the price of fossil fuels started to climb. In Australia, Japan, and Israel, the production of solar water heaters is widespread. In 1960, more than 200,000 solar water heaters were reported to be in use in Japan, with almost the same numbers in domestic installations in Israel and Australia. The reason for the success of solar water heaters in Japan is not only due to the shortage of conventional energy sources, but also to their willingness to adapt their life-style to the solar supply—showering at night instead of in the morning.

Solar water heaters can be constructed in varying degrees of complexity. The simplest design, useful in warm, sunny areas, consists of a black rubber hose sitting out in the sun. By adjusting the rate of flow through the hose, one can get a steady stream of hot water. The hot water can either be used directly, or collected in a storage tank for use later on for dishwashing and showers. Another simple system consists of a shallow trough of water with a transparent cover. This type of solar heater would be filled in the morning, with the hot water drained off for use in the afternoon or early evening.

More complex, higher-quality systems use well-designed, flat-plate collectors in combination with a small, insulated storage tank. A typical system, such as is commonly used in Israel and Japan, is shown in Figure 7-14. A tilted collector is connected to a separate, well-insulated water storage tank by insulated pipes. The bottom of the storage tank is a little over $\frac{1}{2}$ meter higher than the top of the flat-plate collector, with circulation occurring through thermosiphoning. When the water in the collector is heated by the sun, it becomes less dense and rises up to the top of the storage tank. The warm water is replaced by the more dense, cool water coming from the bottom of the storage tank. As long as the sun shines, the water will circulate, and its temperature will rise. Flow in the reverse direction is generally averted because of the height differential between the inlet and outlet ports. Some systems have a check valve that allows water flow in only one direction.

Piping from the collector to the storage tank should be sufficiently large so that thermosiphoning will occur without the need for extreme temperature differences. Typically, a 2.5-cm pipe suffices. The storage tank is commonly large enough to hold a two day's supply of hot water. Standard practice indicates that the average person consumes about 75 to 115 liters per day. A larger storage tank will carry over longer sunless periods, but will cost more. The supply pipe to the bottom of the collector should feed from the bottom (coolest part) of the storage tank. Hot water from the collector should feed almost to the top of the storage tank. The inlet should be sufficiently below the top so that about a half

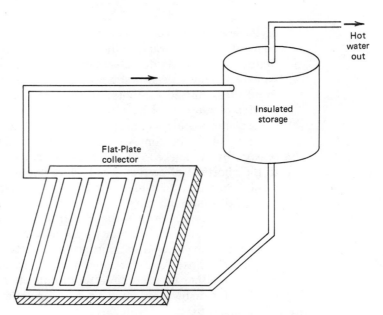

Hot
water
out

Insulated
storage

Flat-Plate
collector

FIGURE 7-14   Illustration of a typical solar hot water heating system operating on the thermosiphon principle.

day's worth of hot water is above the inlet. This is done because water
coming from the collector may not always be as hot as that at the top
of the storage tank because of low-insolation rates.

The flat-plate collector area needed will vary considerably depending
upon local conditions. A rule-of-thumb estimate is that 1 m² of collector
area is required to heat 61 liters of water per day. Thus, if a family of
five uses 380 liters per day, the necessary area of flat-plate collector
would be about 6 m². Most of the commercially successful thermosiphoning water heaters are produced in warm climates. In cold climatic
regions, protection against freezing must be provided. There are three
common ways to do this: (1) Provide movable insulation to cover the
collector; (2) Have a means of draining the collector; (3) Circulate an
antifreeze solution rather than water. This latter solution is the most
common one, but then a heat exchange between the solar heated solution and the water in the storage tank is required. Additionally, this
eliminates the use of thermosiphoning systems; an active system with
a small pump to drive the circulating fluid is now required. This can be
accomplished by running the antifreeze solution through a coil of metal
tubing immersed in the storage tank. Since some antifreeze solutions
such as ethylene glycol are toxic, great care must be taken with the
design. Most housing codes require the use of either two heat exchangers or double-walled heat exchangers when toxic fluids are used,
thereby reducing the overall efficiency of the system.

When the space heating requirements of a home are provided for by a solar system, the large heat storage system can be used as a preheater for a conventional water heater. For most air or water space heating systems, a simple heat exchanger transfers heat from the storage to the water on its way to the regular hot water heater.

BIBLIOGRAPHY

1. H. C. Hottel and D. D. Erway, "Collection of Solar Energy," Chapter 5, in *Introduction to the Utilization of Solar Energy,* Edited by A. A. Zarem and D. D. Erway (New York: McGraw-Hill, 1963).
    For those interested in a fairly comprehensive treatment of flat-plate collectors, this concise article is an excellent place to start. The effect upon the collector efficiency of the materials used, the solar energy input, the number of glazed surfaces, and other parameters are discussed and results presented for typical cases.

2. J. A. Duffie and W. A. Beckman, "Flat-Plate Collectors," Chapter 7, in *Solar Energy Thermal Processes* (New York: Wiley, 1974).
    A detailed description of flat-plate collector design. This book is the "Bible" for many different solar energy applications. Should be on the must reading list for anyone interested in the technical aspects of solar energy.

3. J. Kreider and F. Kreith, *Solar Heating and Cooling* (New York: McGraw-Hill, 1975.)
    Chapter 3 of this recent book contains a great deal of useful technical information.

4. G. Daniels, *Solar Homes and Sun Heating* (New York: Harper & Row, 1976).
    Chapter 6 of this book contains an elementary description of flat-plate collectors along with practical construction details.

5. "Solar Water Heaters," *Popular Science,* 99 (May 1976).
    A well-illustrated article on how to build a solar water heater. The article immediately following describes several commercial heaters that are available now.

6. F. de Winter, *How to Design and Build a Solar Swimming Pool Heater* (New York: Copper Development Association Inc., 1975).
    Contains a very comprehensive review of the principles of operation of flat-plate collectors. A detailed discussion of the economics and construction practices for hot-water heating is presented.

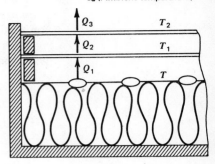

T_a (Ambient temperature)

Q_3 T_2

Q_2 T_1

Q_1 T

PROBLEMS

The figure at the right is drawn
to represent a cross-sectional view
of a typical solar collector.
The next 24 questions refer to this
figure.

1. The collector heat-exchange fluid is usually not which of the fol-
    lowing?
    (a) Air.   (c) Ethylene glycol and water.
    (b) Water.   (d) Oil.
2. The most serious problem with using a thin plastic sheet as the
    glazing material is:
    (a) It is structurally too weak.
    (b) It absorbs too much of the incoming radiation.
    (c) It becomes partly opaque after prolonged exposure to ultravi-
        olet radiation.
    (d) It is too expensive for routine use in collectors.
3. The upward heat loss through the top of the collector is typically
    what percentage of the total collector heat loss?
    (a) 5%   (c) 50%
    (b) 25%   (d) 85%
4. The percent transmission of the incoming solar radiation through
    the top cover plate is typically:
    (a) 98%   (c) 86%
    (b) 94%   (d) 50%
5. The absorptivity of the black absorber plate when covered with a
    common flat-back black paint is typically:
    (a) 80%   (c) 94%
    (b) 85%   (d) 98%
6. The radiative heat loss between the top cover plate and the outside
    world is proportional to:
    (a) $T_1^4 - T_2^4$ (c) $T_2^4 - T_a^4$
    (b) $T_1^4 - T_a^4$ (d) $(T_2^4 - T_1^4)^2$
7. Compare the heat loss between the absorber plate and the cover
    plate with that between the cover plate and the outside world.
    (a) The two heat losses are equal.
    (b) The heat loss between the cover plate and the outside world
        is greater.

(c) The heat loss between the absorber plate and the cover plate is greater.

(d) Insufficient information available to decide.

8. Suppose that a 15-kilometer-per-hour (kmph) wind is blowing. The dominant heat loss mechanism (or mechanisms) between the absorber plate and the adjacent cover plate is:

(a) Convection.   (c) Radiation.

(b) Conduction.   (d) Convection and radiation.

9. The efficiency of this solar collector is largest at which of the following temperatures for the absorber plate? Assume $T_a = 5°C$.

(a) 20°C   (c) 65°C

(b) 50°C   (d) 95°C

10. When solar radiation strikes the *boundary* between the air and the top of the cover plate what percentage of the incident intensity is reflected?

(a) 0%   (c) 8%

(b) 4%   (d) 12%

11. The distance between glazings is typically:

(a) 0.10 cm.   (c) 1.0 cm.

(b) 0.35 cm.   (d) 10 cm.

12. Suppose that the absorber plate is coated with a selective surface. Which of the following far-infrared emissivities would offer the greatest reduction in collector heat losses?

(a) 0.05   (c) 0.50

(b) 0.10   (d) 0.95

13. Suppose that the reflectivity of the black absorber surface is 0.06 for visible light. What is the absorptivity of this same surface in the visible region?

(a) 0.06   (c) 0.94

(b) 0.12   (d) 1/0.06

14. For routine operation of the solar collector at almost 100°C the optimum number of cover plates would be (assume no selective surface):

(a) 0   (c) 2

(b) 1   (d) 5

15. Suppose that the collector absorber plate is made of roll-bond aluminum. Which of the following should be used as the heat-exchange fluid?

(a) Water.

(b) Air.

(c) A mixture of ethylene glycol and water with a corrosion inhibitor.

(d) A mixture of water and finely ground particulate matter.

16. Assume that the black absorber plate has been painted with a

typical type of flat-black paint. About what percentage of the incident solar radiation is reflected back by the absorber plate?

(a) 94%   (c) 6%

(b) 50%   (d) 1%

17. Assume that $T_2 - T_1$ is about 20°C. How does the magnitude of Q_2 compare with that of Q_1?

(a) $Q_2 = \frac{1}{2}Q_1$ (c) $Q_2 = 2Q_1$

(b) $Q_2 = Q_1$ (d) $Q_2 = 10Q_1$

18. Which type of heat loss between the absorber plate and the adjacent cover plate can be neglected?

(a) Conduction.   (c) Insulation.

(b) Convection.   (d) Radiation.

19. Under the equilibrium conditions what can you say about T as compared to T_1?

(a) T is almost equal to T_1 but is slightly higher.

(b) T and T_1 are exactly equal.

(c) T is somewhat less than T_1.

(d) T is somewhat greater than T_1.

20. Suppose that the absorptivity of the black absorber plate was normally about 0.90, and we replaced it with a black absorber plate with $\alpha = 0.94$. Over an entire heating season the performance of the solar collector will:

(a) Decrease by 4%.   (c) Increase by 4%.

(b) Stay the same.   (d) Increase by 2%.

21. Suppose that a 15-kilometer per hour (kmph) wind is blowing. The *dominant* heat loss mechanism (or mechanisms) for heat transfer between the two glazings is:

(a) Convection and radiation.

(b) Convection and conduction.

(c) Convection.

(d) Radiation.

22. Suppose that a 15-kmph wind is blowing. The *dominant* heat loss mechanism (or mechanisms) for heat transfer between the top glazing and the outside world is:

(a) Convection and radiation.

(b) Convection and conduction.

(c) Convection.

(d) Radiation.

23. An ideal solar collector would have the black absorber plate at a constant temperature. In fact, the absorber plate has a strongly varying temperature distribution. This latter situation gives rise to a changed efficiency of the collector over that for an ideal one under any given operating conditions. What can be said about the efficiency of the actual solar collector relative to the ideal one?

(a) Nothing.
(b) It will decrease.
(c) It will increase.
(d) It will remain the same.

24. One of the following is not a good way to improve the performance of a flat-plate solar collector.
(a) Use a selective surface.
(b) Use an antireflective coating.
(c) Use three cover plates.
(d) Use an evacuated collector.

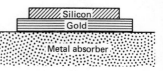

The next four questions refer to the picture of one type of selective surface (shown at the right).

25. The incident solar radiation is absorbed by:
(a) The layer of silicon.
(b) The layer of gold.
(c) The metal absorber plate.
(d) The water in the pipes in contact with the absorber plate.

26. If infrared radiation were incident on the system it would be reflected by:
(a) The layer of silicon.
(b) The layer of gold.
(c) The metal absorber plate.
(d) The water in the pipes in contact with the absorber plate.

27. The long wavelength radiation emitted by the hot absorber plate:
(a) Passes through the gold and silicon without any appreciable loss.
(b) Is mostly reflected back by the layer of silicon.
(c) Is mostly reflected back by the layer of gold.
(d) Is absorbed by the layer of silicon.

28. The overall emissivity for long-wavelength radiation of this selective surface "sandwich" is typically:
(a) 0.95   (c) 0.10
(b) 0.50   (d) 1/0.10

29. Which of the following is a major problem with present commercial selective surfaces?
(a) Their selective emissivity property tends to degrade when operated in the air for a few years.
(b) They are very inexpensive.
(c) They will not adhere properly to the metal absorber plate.
(d) Their absorptance for visible light is too low.

30. Evacuated flat-plate collectors are better than nonevacuated ones at high operating temperatures because:
    (a) Convective heat losses, as from the absorber plate, are greatly reduced.
    (b) Conductive heat losses from the absorber plate are greatly reduced.
    (c) Radiative heat losses from the absorber plate are greatly reduced.
    (d) Convective heat losses from the top cover plate are greatly reduced.

31. In the Thomason type of flat-plate collector the heat exchange fluid:
    (a) Is air.
    (b) Flows through pipes in contact with the absorber plate.
    (c) Flows under gravity down depressions (corrugations) in the absorber plate.
    (d) Is propylene glycol.

32. A severe disadvantage of using a mixture of ethylene glycol and water as the heat exchange fluid for a flat-plate collector is that:
    (a) It won't freeze at $-1°C$.
    (b) It is toxic.
    (c) It has an extremely low specific heat.
    (d) Too much pumping power is required.

33. An advantage of using copper for the absorber plate is that:
    (a) It has a higher thermal conductivity than other metals.
    (b) It is cheaper than steel or aluminum.
    (c) It will not corrode easily so that water can be used as the heat-exchange fluid.
    (d) It weighs less than steel or aluminum.

The efficiency of different flat-plate collectors is pictured in the figure to the right. The next five questions refer to this figure.

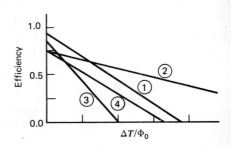

34. Which collector has the largest heat loss coefficient?
    (a) 1   (c) 3
    (b) 2   (d) 4

35. Which collector is best at high operating temperatures?
    (a) 1   (c) 3
    (b) 2   (d) 4

36. Which collector has the least loss of the solar intensity through the cover plates?

(a) 1   **(c)** 3

(b) 2   **(d)** 4

37. One of the four collectors is evacuated (has a vacuum). Which one is it?

(a) 1   **(c)** 3

(b) 2   **(d)** 4

38. Which collector is best for low operating temperatures?

(a) 1   **(c)** 3

(b) 2   **(d)** 4

39. One of the following is not a good way to provide freeze protection for a flat-plate collector system.

(a) Use movable insulation around the collector.

(b) Circulate an antifreeze solution.

(c) Drain the collector when it is not operating during periods of low ambient temperatures.

(d) Increase the rate of water flow through the collector by a factor of 2, leaving it on at all times.

40. Water circulates in a thermosiphon, hot-water heating system by:

(a) Forced circulation using a pump.

(b) Gravity flow from the top of the collector to the lower storage tank.

(c) The fact that hot water is less dense than cold water.

(d) The fact that cold water is less dense than hot water.

41. A good selective surface will be designed in the following way.

(a) Utilize materials that strongly reflect visible light and absorb the far-infrared radiation.

(b) Utilize materials that strongly reflect the far-infrared radiation and absorb the visible light.

(c) Utilize materials that strongly reflect both the far-infrared and the visible wavelengths.

(d) Utilize materials that strongly absorb both the far-infrared and the visible wavelengths.

42. To the right is a typical cross-sectional view of a liquid-type, solar thermal collector. In the following, a number of comments are made about the solar collector. You are to match the comments on the left with the *best choice* from the list to the right.

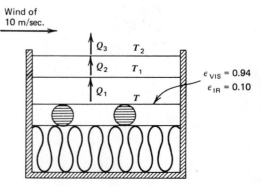

_____ 1. The radiative portion of the heat loss Q_1 is

_____ 2. The dominant heat loss mechanism for Q_3 is

_____ 3. The amount of sunlight passing through the first glass cover is

_____ 4. The amount of sunlight absorbed by the black surface is

_____ 5. The peak wavelength of the radiation emitted by the black surface is roughly

_____ 6. The peak wavelength of the incident sunlight is

_____ 7. At 45°N latitude the best tilt angle from the horizontal for the collector is

_____ 8. Q_1 is _____ T_2

_____ 9. The far-infrared emissivity of the black surface is

(a) greater than
(b) 94%
(c) 10,000 nm
(d) 30°
(e) convection
(f) proportional to $T_1 - T_2$
(g) less than
(h) equal to
(i) proportional to $(T - T_1)^2$
(j) proportional to $(T - T_1)$
(k) 0.1
(l) 0.9
(m) 55°
(n) 2000 nm
(o) 500 nm
(p) 84%

Space Heating and Cooling

VIII

In this chapter we take up the study of the basic components of active solar systems—systems that use energy in addition to solar energy for their operation. Passive systems, which achieve the transfer of thermal energy through natural means, are discussed in the next chapter. The principal elements of a complete active solar heating system are illustrated in Figure 8-1. The solar energy is captured using some type of a flat-plate collector operating at rather low temperatures. In the collector, either a liquid solution, or air, is circulated to transfer the heat from the collector to a heat storage facility. The heat storage unit may consist of either a large tank of water, a bin containing many tons of small pebbles, or phase-change materials in containers. Heat is then delivered to the house either directly from the collectors or from the

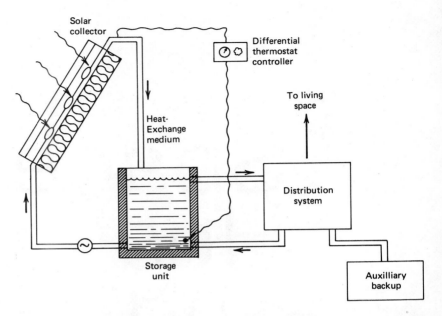

FIGURE 8-1   Schematic drawing of a complete active solar heating system. See text for more details. Provision for preheating the hot-water supply may also be included.

heat storage unit. A system of controls is needed to regulate the delivery of heat to the house and to insure efficient operation of the solar collector. An auxiliary backup heating system is generally necessary to provide heating during extensive periods of low solar radiation. Some solar heating systems may include a heat pump. The solar system can also be adapted to provide air conditioning, if desired.

Before going on to a detailed discussion of the key solar system components, a few general comments about the siting and construction of the solar systems are in order. It should be kept in mind that not every building site is suitable for good solar energy collection. It is imperative to locate the site so that an adjacent building will not appreciably shade the solar house or solar collector. This might mean placing the house as near to the north side of the lot as zoning regulations permit. Deciduous trees planted east and west of south-facing windows are advantageous for providing summer shade, but should not block off the winter sun.

Designers of solar houses have concentrated on building energy-efficient structures. The use of large amounts of insulation follows this discussion, but there are also a large number of other worthwhile items to consider. Spacious high ceilings of the cathedral type waste heat and should be avoided if possible. Careful fenestration design can also be of great benefit. A normal window area on the south side of the house can easily provide most of the daytime-heating requirements of a well-insulated house. Too much glass space can create uncomfortably warm, inside temperatures, even in the middle of winter, unless materials such as masonry are used within the structure, which can absorb and store heat. Windows on the north side should be minimized as they let in the diffuse light. Serious heat loss at night can be avoided with the aid of various means of insulation. It is preferable to double-glaze all windows. Small, screened openings near the floor and ceiling that can be opened and closed give you better ventilation than an open window, since natural convection is assisted. All exterior doors should be insulated whenever possible and should have weather stripping.

House insulation is always a wise investment; insulation is cheap compared to the cost of the solar heating system necessary to overcome heat losses. Sidewalls should contain at least full width, 9-cm fiberglass batt insulation. Polystyrene foam sheathing may be used instead of the typical impregnated board on the outside of the walls. Other good construction practices such as weather stripping and caulking should be emphasized. Home insulation is usually specified in the terms of ''R'' values. In mathematical terms, the R value is the reciprocal of the heat transfer coefficient, $U(R = 1/U)$. For good house insulation one should have as low a heat transfer coefficient as possible, which means a high R value. A few examples will help clarify this discussion. The R value of a 7.5-cm, hardwood door is about 2.7 in English units where the units of U are Btu/hr-ft^2-°F. By contrast, 5 cm of polystyrene insulation has an R value of about 8, and 9 cm of fiberglass insulation has an R value of 11. This latter is a good minimum target value for the wall design of a new house. Better yet, going to 5-cm by 15-cm (2 in. × 6 in.) studs will permit increasing the insulation to R-19.

A.
THE SOLAR
COLLECTOR

In Chapter VII the important elements in the detailed design of efficient flat-plate collectors were covered. Now we consider the important practical factors in the choice of the collector to be used in the complete active solar heating system. These factors include the decision as to whether air or a liquid will be used as the heat-transfer medium, the area of collector needed, the economic cost involved, and whether a commercial or home-made collector is to be installed. Most collectors are located on the roofs of buildings, but consideration should be given to a location at ground level. The utilization of specular reflectors to enhance the light gathered is discussed later.

Determining the area of collector needed depends upon both climatological and architectural considerations. A common method used to estimate the required collector area is to use the concept of degree days

(DD). This is defined as the difference between the inside and outside temperatures of a house. Standard practice uses an indoor temperature of 65°F as the base from which to measure DD, since most buildings do not require heat until the outdoor temperature is between 60 and 65°. The outside temperature for a given day is then defined as the arithmetic mean of the minimum and maximum temperature. If the minimum and maximum temperatures during a given day were 42 and 56°, respectively, then the number of DD would be $65 - 49 = 16$. Typical heat loss ratings for houses range from 15 to 40 million J/DD. For example, consider a house with a 15 million J/DD rating located in a climatic area with a heating load of about 4000 DD per year (Seattle, Washington). The total space heating requirement would then be $15 \times 10^6 \times 4000 = 60 \times 10^9$ J/yr. (This is about 16,700 kW-hr/yr.)

If we know the annual heat load of the house, it is then possible to estimate the area of solar collector needed. Usually, one does not choose a collector area sufficient to provide the entire heating load (including hot water) of the house. A large number of studies have been made to determine the optimum size. The answers vary from one study to the next, depending on such factors as climatic conditions and the projected cost of nonsolar energy. While it is impossible to give a precise answer to this question, it is probably safe to say that the solar collector should provide about 50 to 70% of the total annual heating load. Clearly, the system should not be designed to provide the total heating requirements for the worst winter months of December and January, since then an excess amount of useful heat would be provided in the other months.

The choice of heat-exchange fluid depends upon local climatic conditions, the type of collector chosen, and the individual preference of the user. Air systems have the advantage that they do not freeze up, and no corrosion of the collector is expected. On the other hand, a very air-tight circulation system is required, considerable electrical power is needed for moving the air by fans or blowers, and the efficiency of an air-type, flat-plate collector is somewhat below that achieved with liquid systems. Designers of air systems note, however, that the overall efficiency for delivering useful heat to the building is the same for liquid and air systems because of a better extraction efficiency from the heat storage medium when air is used. Liquid systems have the advantage of ease of handling and high collector efficiency. They require protection against freezing, and many of the common materials used for the absorber plate (copper excepted) are subject to corrosion unless precautionary methods are taken.

The solar collector is usually tilted in order to enhance the collection of the direct component of the solar radiation. A rule-of-thumb estimate for the tilt angle for winter heating optimization is to orient the collector at an angle equal to the latitude plus 10 to 15°. This is born out by the results of a study by the Los Alamos solar energy group, which is shown in Figure 8-2. At a latitude of 42° the optimum collector tilt angle is between 50 and 62°. The calculations were performed for solar systems that could provide either 40 or 75% of the total space heating load. The performance is not very sensitive to the orientation angle, being within 2% of the maximum for tilt angles covering the range of 38 to 65°. It is also of interest to study the azimuthal sensitivity of solar energy collections as presented in Figure 8-3. Although the maximum performance is obtained when the collector is oriented due south, a variation of 30° east or west only reduced the performance by about 2.5 to 5%. This conclusion is a function of the climatic region, however, and an orientation of 15 to 20° west of south will often be favored for two reasons. First, many areas often experience haze or fog in the early morning hours, which will cut down on the incident solar intensity. Furthermore, higher ambient temperatures in the afternoon will raise the collector efficiency for this period. A good rule for most climatic regions is to orient the collector so as to point directly toward the sun at 1:00 P.M. (solar time).

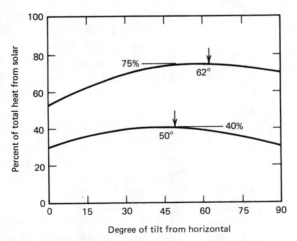

FIGURE 8-2 The amount of useful heat collected by a solar system over the winter heating season in Medford, Oregon (42° North latitude) has been calculated by the Los Alamos solar group as a function of the collector tilt angle. The two curves represent calculations for systems capable of providing 40 and 75% of the total heating load from solar energy when optimally oriented. Small variations from the optimum result in only a very minor reduction in the solar energy collected.

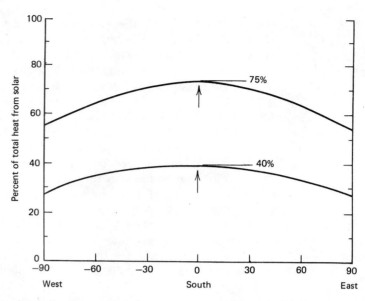

FIGURE 8-3   Calculations similar to those for Figure 8-2 show that a due south orientation is nearly optimum. For exceptions to this rule see the text. The decrease in performance for variations from due south is somewhat greater for colder northern cities than for warmer ones in the southern part of the country.

**B.
HEAT STORAGE**

The size of the solar collector is also influenced by the size of its associated storage system. Since space heating tends to be out of phase with the sun, both daily and annually, storage of the solar energy is required, with the amount again dictated by local conditions. In many areas, one to three days storage is sufficient, while in cloudy areas greater storage capacity may be warranted. Theoretically, sufficient storage capacity could be provided to cover the accumulation of fall heat for winter use. For climatic areas where cooling is the dominant requirement the amount of storage capacity may be greatly reduced, since the cooling need is essentially in phase with the sun.

Storage of solar heat is commonly accomplished with gravel beds, water tanks, or with heat-of-fusion salts where the principal mechanism for heat storage comes from the change of phase of the salt from a solid to a liquid. (The use of wet dirt has also been reported.) The storage medium naturally depends upon the type of flat-plate collector used. If a liquid is used to collect heat, a large water tank is the natural choice. If air is used to collect heat, a bed of gravel should be the storage medium. Heat-of-fusion salts can potentially store the most heat per unit volume, but their use has been limited by problems associated with extracting the stored heat and by cost.

The heat storage properties of some common materials are shown in Table 8-1. The heat storage ability of a substance is measured by its specific heat, which was defined in Chapter II as the amount of heat in calories required to raise the temperature of 1 gram of a substance by 1°C. Water can store more heat than most materials because of its larger specific heat capacity. While water has a specific heat of 1 cal/g-°C, the specific heat of rock is only about 0.20 cal/g-°C, a value that is typical of most other solid materials such as iron and aluminum. The common antifreeze, ethylene glycol, has a specific heat of 0.6 cal/g-°C. For a substance that changes phase the important parameter is the amount of heat required per kilogram (or per unit volume) to cause the change from a solid to a liquid at the melting point of the material. The last item in Table 8-1 is for a typical heat-of-fusion salt, sodium sulfate decahydrate, which changes phase from a solid to a liquid at 32°C.

Water is the most commonly used storage medium, since it is the practical choice when a fluid is used as the heat-exchange medium for the solar collector. Let us illustrate its use as a storage medium by calculating the amount of useful heat that can be stored in a 11,370-liter tank (3000 gal) that contains water at a uniform temperature of 54°C (130°F). Assume that useful heat can be withdrawn down to about 27°C (80°F), giving a temperature rise of 27°C in the storage medium. The heat, Q_s, stored in the storage medium is

$$Q_s = 11{,}370 \text{ liters} \times 10^3 \text{ g/liter} \times 1 \text{ cal/g-°C} \times 27\text{°C}$$
$$= 307 \times 10^6 \text{ cal}$$
$$\cong 1.28 \times 10^9 \text{ J}$$

This is the equivalent to two to three days of the winter heating load in a mild climate.

Table 8-1   Heat Storage Properties of Common Materials

Medium	Specific Heat	
	Cal/g-°C	Cal/m³-°C
Water	1.0	1000×10^3
Water and ethylene glycol (50-50 mix)	0.83	872×10^3
Gravel	0.20	278×10^3
Sodium sulfate decahydrate	50 cal/g	83×10^6 cal/m³

A system for heat storage using water as both the heat-exchange fluid and the storage medium is illustrated in Figure 8-4. Water is drawn from the bottom of the storage tank, pumped through the collector, and returned to the top of the storage tank. When the system is not running, air (or an inert gas) enters into the collector and piping, and the water drains into the storage tank. Very often a fluid other than water (such as water plus corrosion inhibitors) is used as the heat exchange medium. In this case the heat transfer liquid is pumped through the collector, then through a heat exchanger in the storage tank, and then back to the collector in a closed loop as shown in Figure 8-5. Since the loop is always filled, it must be protected somehow, from freezing. The storage tank walls should be carefully insulated in order to minimize heat losses. Rather than sheathing the tank with an insulating material, the water tank is often surrounded with small rocks. This helps insulate the tank from the neighboring environment while at the same time providing additional storage capacity. Some disadvantages of water storage systems include the possibility of leaks, the cost of the tank, and the rather large space and weight requirements.

Crushed rock or gravel is used as the storage medium when air is used for heat transfer. The heat capacity per unit weight of rock or gravel is only one-fifth that of water, and the heat capacity per unit volume of rock or gravel is about one-third that of water (see Table 8-1). Thus considerably more rock is required to store the same amount of heat stored in a given amount of water. In the example above, where

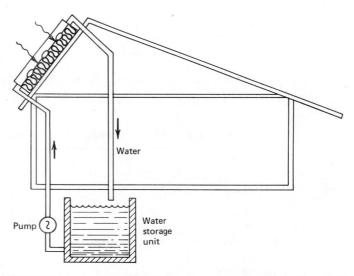

FIGURE 8-4   Schematic drawing of an open-circuit, liquid collector system using water as the heat storage material.

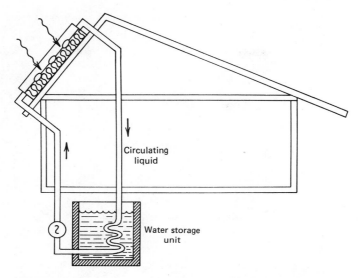

Circulating
liquid

Water storage
unit

FIGURE 8-5 Often a heat transfer liquid other than water is used
to transfer heat from the solar collector to the water
storage unit. In this case some type of heat exchange,
perhaps with a coil, is needed to transfer the heat
from the circulating liquid to the water storage.

307 million cal of sensible heat (the energy stored by raising the temperature of a substance) were stored by raising the temperature of
11,370 liters of water 27°C, to store the same amount of heat would
require three times this volume of rock.

When rock is used for heat storage, only small rocks about the size
of a fist or smaller should be used. This is necessary to increase the
amount of surface area of the rock. The ability of rock to exchange heat
with the circulating air is a function of its surface area per unit volume.
Thus, a cubic foot of small rocks will have much better heat transfer
properties than a solid rock of the same size. In packing the smaller
rocks randomly into the cubic foot of space, only about one-half as
much material by weight will be contained compared to the situation
with a cubic foot of solid rock. This means the total heat storage
capacity is reduced by this factor of $\frac{1}{2}$, but the large gain in the ability
to exchange heat with the circulating air is the most important consideration.

A practical solar air system with rock-bed heat storage is shown in
Figure 8-6. The rocks are contained in a cylinder, and their weight is
supported by a steel grill at the bottom of the container. The central
pipe allows the rocks to be shunted out, when the heat is needed,

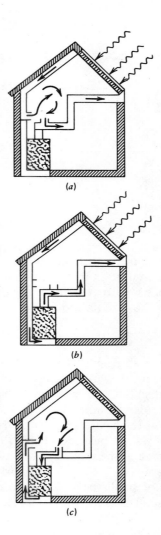

FIGURE 8-6   Illustration of the operating principles of an air-type solar system. (a) The heat collected by the solar collector is transferred directly to the building. (b) The collected heat is stored in the rock bin. (c) Shows how heat is removed from the rock storage and transferred to the building. This type of system is similar to the one manufactured by Solaron, Inc. of Denver, Colorado.

directly to the structure. Rock piles have the advantage that they can neither freeze nor leak. An important aspect of rock storage is the stratification of temperature that is formed. The heated air coming from the collector gives up heat as it passes through the rock storage, exiting as cool air. This results in the entrance portion of the storage being at a higher temperature than the exit portion. In the evening, when the collector is off and the building needs heat, the direction of air flow through the rocks is reversed so that the air leaves from the hottest part of the rock storage system. This results in air to the building that is considerably warmer than if the rock storage were at a uniform temperature. The net result of this temperature stratification is an improved efficiency for the transfer of heat from the storage system to the residence over that obtained with liquid solar systems. This explains the comment made earlier that the overall efficiencies of air and liquid systems are about the same even though the collector efficiency for the liquid system is higher.

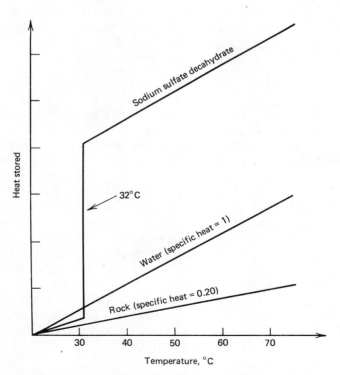

FIGURE 8-7   Comparison of the heat capacities of various storage media. Notice the large amount of heat that can be stored in the heat-of-fusion salt.

The greatest advantage of storing heat with phase-change materials is illustrated in Figure 8-7. A large amount of heat per unit weight is required to effect the melting of a solid as compared to that needed to raise its temperature 1°C. As seen in Figure 8-7, once sufficient heat is added to cause the entire storage medium to melt, further heat is stored as sensible heat just as is the case for other materials. Thus, heat-of-fusion salts can be used to store large amounts of heat in small volumes of materials. Commercial development of this storage approach has been slow because of problems arising from their behavior after having been thermally cycled a number of times. Sometimes a portion of this phase-change material sinks when frozen, and it is not a simple matter to return to a homogeneous fluid again upon melting. In addition, phase-change materials have a tendency to supercool, whereby the storage material will fail to solidify at the freezing point. Another problem is associated with the failure of containers after prolonged use. A current approach to overcome some of the difficulties has been to use long, 2.5-cm thick trays to store the phase-change materials. Much progress has also been made in developing chemicals to overcome the problems of supercooling and segregation. Because of the attractiveness of storing a large amount of energy in a small volume, it is certain that research and development in this area will continue until reliable, economical systems have been achieved.

C. DELIVERY AND CONTROL SYSTEMS

Once heat has been collected and stored it must still be delivered to the building in order to be useful. The delivery system must extract heat from the storage system and deliver it to the house. The controls for this can be thermostats operating in much the same way as a conventional unit. There will be some differences, however. For example, in the case of the air system shown in Figure 8-6, it is possible to deliver heat directly to the house from the solar collector. At other times the control system will switch the arrangement to permit the flow of warm air from the storage unit to the building. The control system allows returning air from the building to go directly to the solar collector during the day (unless cloudy) and to go to the cool end of the rock storage system at night.

One way to extract heat from the water storage system for delivery to the house is to supply the house with hot water from the top of the storage tank and to withdraw fluid from the cooler bottom for delivery to the collector. This results in the warmest fluid being used for space heating while the collector is supplied with the coolest fluid, resulting in maximum collector efficiency. Water from the storage system can also be circulated directly to the building and passed through coils as illustrated in Figure 8-8.

The overall arrangement of a solar heating system is shown schematically in Figure 8-9. For an air-type solar collector, the key controls are the house thermostats and the differential thermostat controller. When the house thermostat requires heat, it directs the warm air into the house heating system. The differential thermostat controller regulates the flow of air to the collector and back to the storage tank. Temperature-sensing devices are used at the output of the solar collector as well as at the cool end of the storage tank. The fan that circulates air through the solar collector is turned on when the temperature at the collector outlet is somewhat greater than that of the temperature of the cool part of the tank. The operation of the differential control system will result in a pattern much like that shown in Figure 8-10 for a collector using a liquid heat transfer medium.

In the cooler portions of the United States a full-capacity auxiliary heating unit must be provided to supplement the solar heating system when no useful solar energy is being collected and when all the stored heat has been consumed. The auxiliary furnace is usually arranged to heat the heat transfer medium and distribute it through the same channels as employed for solar heat. This can be done in a variety of ways. Air from the solar collector can be passed through the furnace, if necessary, before circulation to the rooms. Sometimes auxiliary heat is provided to the heat-storage system. For water systems another ap-

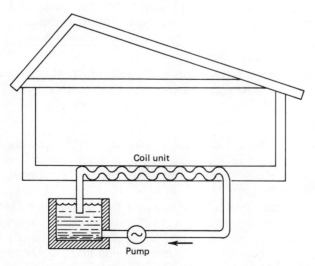

FIGURE 8-8   Schematic illustration of the method of radiant heat distribution using a coil unit in the floor. The coils could also be located in the walls or ceiling of the building.

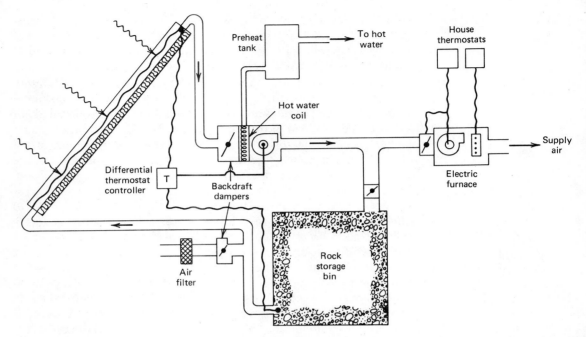

FIGURE 8-9 Possible arrangement for solar space and hot water heating using an air collector and a rock storage unit. Additional controls, not shown, are needed for the hot water, preheating system.

proach is to heat water in a small, auxiliary water tank, from which it is directed either to the house or to the air-flow heat transfer system.

D.
SOLAR COOLING

The use of solar energy to provide cooling is attractive because the cooling is needed when the solar energy is most available. If the local climate requires both heating and air conditioning, a combined system to provide for both of these building needs may prove particularly economical because of a better overall system-usage factor. Although at present only a small fraction of the energy utilized for space heating and cooling goes to meet the latter need, the number of buildings incorporating air conditioning units has been steadily increasing. From the point of view of conservation of energy it makes good sense to use the energy incident upon a structure to provide the required cooling.

In this section we discuss various methods of obtaining solar assisted cooling. Before doing this it is worthwhile to point out a few general conclusions relevant to the solar collectors used for combined systems, providing both heating and air conditioning. First, the optimum collector tilt angle is nearly equal to the latitude angle. Also, the optimum collector area needed is always larger than the area that would be used by a system providing space heating only. Finally, the solar collector should operate at temperatures considerably higher than those required for heating. These higher operating temperatures imply the use of low-heat-loss collectors. One way to achieve this is with evacuated collectors; another is to use a triple-glazed collector.

Schemes to provide solar cooling can be divided into three broad categories:

1. *Systems that provide refrigeration.* The cooling can be provided by a system using absorption refrigeration, or by a conventional vapor-compression, air conditioning unit where the needed mechanical energy is derived from the solar input.

2. *Dehumidification systems.* Often the removal of moisture from the air in the living space is desirable. This approach is typified by desiccant systems in which air is dried and then cooled by evapora-

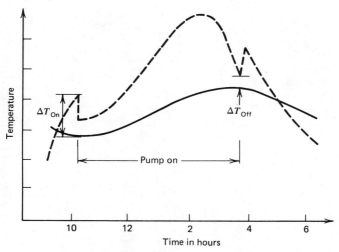

FIGURE 8-10   Illustration of how the differential thermostat controller is used to turn the solar collector pump on and off (for a liquid system). The solid line depicts the temperatures at the bottom of the storage tank, while the dashed line shows the temperature at the collector outlet. For liquid systems, ΔT_{on} is about 11°C and ΔT_{off} is about 3°C.

tion. Solar energy is used as the source of heat for regenerating the air-dehumidifying agent.

3. *Natural cooling methods.*   Heat is given off to the night air and sky by convection and radiation. The storage unit is utilized during the day for absorption of heat from the living space and the stored heat given off at night.

The most common way to achieve solar cooling is to use absorption refrigeration. Thus far, only absorption cycles using lithium bromide and water, and ammonia and water have obtained any commercial success. The arrangement of a typical absorption cooling unit is shown in Figure 8-11. In this approach pressurization is accomplished by dissolving the refrigerant in a liquid in the absorber section. The solution is then pumped to a higher pressure. The low-boiling-point refrigerant is then driven from the solution by adding solar heat in the generator. The vapor then passes on to the condenser where heat is given off. The

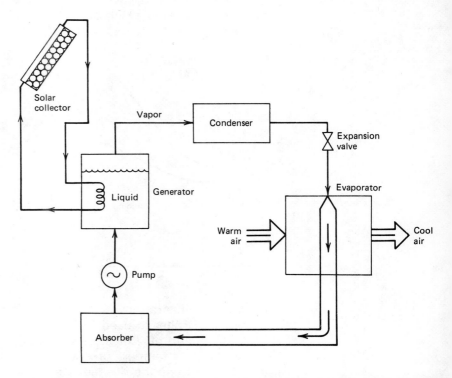

FIGURE 8-11   Flow diagram of a heat-actuated solar refrigeration system.

condensed refrigerant passes on to the evaporator through the expansion valve. In the process of evaporation, heat is extracted from the air to provide the cool air (or liquid in some cases) needed for cooling. Finally, the refrigerant vapor passes back to the absorber where it is absorbed back into the solution.

In the ammonia-water, heat-actuated refrigeration system, water is the absorber. At low temperatures water can absorb a large quantity of ammonia. When heated at high pressure the ammonia is given off as a gas. The ammonia can then be condensed back to liquid form and allowed to expand through a valve. Some of the ammonia will evaporate and cool down the remainder. The only external mechanical work needed is that required to raise the pressure of the ammonia-water system. The chilled water resulting from the process is used to provide the desired air conditioning. The lithium-bromide system uses water as the refrigerant. Lithium-bromide will absorb large quantities of water when cool and give it up as water vapor when heated. Regardless of the type of heat-actuated refrigeration process used, operating temperature is a trade-off when coupling a solar collector to a heat-actuated refrigeration unit. This is due to the fact that the efficiency of the solar collector decreases with temperature, while the performance of the refrigeration unit increases with generator temperature. One commercial lithium-bromide system delivers four times the amount of cooling at 104°C than it does at 82°C. One of the key problems associated with absorption cooling systems is associated with the start-up transient. Cooling does not begin immediately when the unit is turned on, but instead starts only after the generator has reached a certain minimum temperature. In practical systems a certain amount of time is required to heat up the generator to the point of maximum efficiency. For systems in which the cooling unit is alternately on and off, this time delay results in an average performance considerably below the optimum.

Considerable attention has been given to methods whereby thermal energy obtained from the sun would be converted to some form of mechanical output. This in turn, would be used to drive a more conventional vapor-compression refrigeration unit, which then performs the function of cooling the building. Compare this method with the absorption equipment described earlier where the thermal energy input directly produces the cooling function. Determining which of these cooling approaches will be best for various regions of the country must await further technological development.

Systems in which moisture is removed from the air through the use of desiccants are most suitable for air systems. In this approach, air-drying agents that can be regenerated by heat are used. Their usefulness is primarily for hot, humid climates where a large portion of the cooling load is for the removal of moisture from the air. The solar energy is then used as a source of heat for the regeneration of the air-dehumidifying agent. A common commercial unit employs triethylene glycol as the dehumidifying agent. Dehumidification is accomplished by alternately absorbing the moisture from the air and then removing it from the resultant solution of desiccant and water by using solar heat.

Given suitable climatic conditions, natural cooling methods have been used with some success. One approach utilizes the fact that a heated surface will radiate energy to the sky under suitable conditions. Heat is transferred during the day from the warm house to the storage medium. Air or water can then be circulated from the storage unit through a large radiating surface, dissipating heat to the sky. This approach works best in the southwestern region of the United States, where there are large differences between day and night temperatures.

Evaporative cooling is obtained through the loss of heat that occurs when moisture is evaporated into the air. The process of passing air through pads that are saturated with water and evaporating some of the moisture picked up by the air has been used for many years as a means of reducing the temperature of desert air to a tolerable level. The problem here is the subsequent rise in relative humidity.

A modification of the simple evaporative cooler has been developed

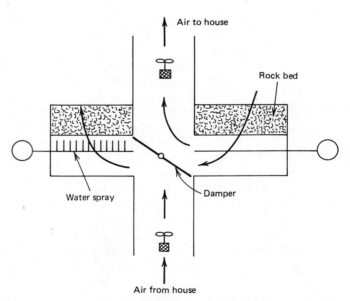

FIGURE 8-12 The Australian rock bed regenerator system provides cooling using the evaporation of water. (Courtesy of J. I. Yellot, "Solar Radiation and its Uses on Earth," published in *Energy Primer*, pp. 4–24, Portola Institute, Menlo Park, Calif.)

in Australia and given the name, "Rock Bed Regenerative Air Cooler." A schematic example of such a system is shown in Figure 8-12. Two beds of rocks are used, which are separated by an air space in which a damper is located. At the beginning of a cycle, water is sprayed onto one of the rock beds for 10 to 15 seconds, thoroughly saturating the rocks. The air from the house evaporates the moisture, cooling the rocks on one side. On the other side, cool air is obtained by drawing air into the house over the previously cooled rock bed. Only a very small amount of moisture is added to this air, since it is the rocks that have been cooled and the only moisture remaining is that small amount adhering to the rocks' surface. The advantage of this technique is that the evaporation of 1.5 gallons of water is equivalent to about 1 ton of refrigeration. The amount of power needed is only about 10% that of a normal refrigeration system. Unfortunately, this system will not work when the outside humidity is high.

**E.
REFLECTOR
ENHANCEMENT**

In areas where there is low insolation during the winter, a standard solar collector may only be turned on to collect useful energy for a few days each month. This is the case for cloudy areas of the country such as the Pacific Northwest. Inexpensive reflectors can be used to enhance the light collection of flat-plate collectors, thereby greatly improving their overall performance.

The concept of using reflectors with flat-plate collectors is not new, having been used as early as 1911 by F. Shuman in conjunction with a tiltable collector array for a solar power plant at Tacomy, Pennsylvania. A modest reflector was used in one of the early MIT houses and also in the solar house designs of H. E. Thomason. The first person to make use of a large reflector in an optimally oriented reflector-collector system was H. Mathew of Coos Bay, Oregon. The striking feature of this house is the nearly vertical inclination of the collector, rather than the customary tilt angle of 55 to 60° for a collector located at 45°N latitude. A description of this house is given in Chapter IX.

A qualitative understanding of why a nearly vertical collector is best when a reflector is utilized can be obtained with the help of the diagrams in Figure 8-13. We assume that the sun's rays are incident at some angle θ (from the vertical). This angle varies with the time of day and year, but detailed calculations show that this qualitative picture is valid. The top diagram shows that the optimum geometry for a simple collector has the incident beam radiation striking the collector normally. In this case the collector tilt angle θ_T equals the angle of incidence θ_i. The second diagram shows that for the reflected beam to strike the collector, the collector tilt angle must be greater than θ_i (for a horizontal reflector). The third diagram illustrates the fact that only a finite reflector length

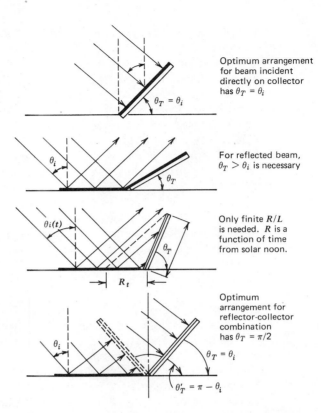

Optimum arrangement for beam incident directly on collector has $\theta_T = \theta_i$

For reflected beam, $\theta_T > \theta_i$ is necessary

Only finite R/L is needed. R is a function of time from solar noon.

Optimum arrangement for reflector-collector combination has $\theta_T = \pi/2$

FIGURE 8-13 Illustration of the reflector-collector geometry showing why a near vertical collector is the optimum arrangement.

is needed for any given angle of elevation of the sun above the horizontal plane. The bottom drawing shows that the collector orientation for optimum gathering of the reflected intensity only occurs when $\theta_T = \pi - \theta_i$. If the reflected intensity and that directly incident upon the collector are roughly equal, then the best collector orientation is the average of the optimum angles for the two individual contributions, leading to the conclusion that the collector should be oriented vertically.

Detailed analysis shows that the optimum reflector orientation at 45°N latitude is about 5 to 10° downward, with an angle of 100° between the reflector and collector planes. In this geometry, the amount of light gathered by the reflector system in the winter is about 55 to 60% greater than that collected by a standard flat-plate collector oriented at 60°.

How large should the collector be? For flexibility in architectural design, the shortest possible reflector length is desirable. Detailed calculations have been made for the case where the reflector was oriented 5° downward and the collector was oriented 85° upward from the horizontal plane. This analysis shows that the amount of light gathered by a system with a reflector length R, equal to twice the collector height L, is only about 4% less than that obtained with an infinite reflector. These same calculations show that losses, due to having a reflector with a finite width, can be kept below 3% (comparing again with an infinite than 3 if the reflector width to collector height ratio (W/L) is larger than 3. (The reflector width is always assumed to be the same as the width of the collector.)

BIBLIOGRAPHY

1. J. I. Yellot, "Solar Radiation and Its Uses on Earth," in *Energy Primer* (Menlo Park, Calif.: Portola Institute, 1974), pp. 4–17.

    An excellent treatment of solar heating and cooling systems at a level that is understandable to the nonspecialist.
2. G. O. G. Löf, "The Heating and Cooling of Buildings with Solar Energy," Chapter 11, in *Introduction to the Utilization of Solar Energy,* edited by A. M. Zarem and D. D. Erway (New York: McGraw-Hill, 1963).

    Discusses the general aspects of solar heating and cooling in a nontechnical manner. Heat exchange at the collector, types of storage systems, delivery of heat to the building, control systems, backup heating, and solar cooling possibilities are reviewed as of 1963. Includes a description of several experimental homes.
3. D. Watson, *Designing and Building a Solar House* (Charlotte, Vt.: Garden Way Publishing Co., 1977).

    Chapters 2, 4, and 5 of this book contain a lot of material relevant to the subject matter of this chapter. The presentation stresses the architectural features of solar heating.

PROBLEMS

1. Suppose that hot water is stored at a temperature that is 20°C higher than room temperature. Since 1 liter of water is about 1000 cm³, about how many liters of water are needed to store 20 million cal of heat? (Remember that 1 cm³ of water weighs about 1 gram.)
    (a) 100 liters
    (b) 200 liters
    (c) 1000 liters
    (d) 2000 liters
2. By roughly what factor would you expect the rate of heat loss of a house to decrease if the thickness of insulation in the walls and ceiling is doubled?
    (a) None
    (b) 2
    (c) 4
    (d) 8

3. Consider normal winter operation of a solar collector at latitude 45°. The optimum tilt angle (from the horizontal) is near:
   (a) 45°   (c) 60°
   (b) 0°   (d) 90°

4. Suppose that the solar intensity in your climatic area were composed entirely of diffuse radiation. Your latitude is 50°. What would be the optimum tilt angle (from the horizontal) for your collector?
   (a) 45°   (c) 60°
   (b) 0°   (d) 90°

5. For a 20°C rise in temperature, which of the following materials will store the most heat?
   (a) A ton of rocks.   (c) A ton of water.
   (b) A ton of wood.   (d) A ton of iron.

6. The performance of a heat-actuated, solar cooling system is most efficient at which of the following average temperatures for the absorber plate? Assume an ambient outside temperature of 25°C.
   (a) 15°C   (c) 70°C
   (b) 45°C   (d) 100°C

7. In a climate where there is a great deal of morning fog, how should the solar collector be oriented?
   (a) Directly south.   (c) 30° East of South
   (b) 15° East of South.   (d) 15° West of South.

The figure at the right illustrates the behavior of four different types of heat storage materials. The following three questions refer to this figure.

8. Which material has the largest specific heat?
   (a) 1   (c) 3
   (b) 2   (d) 4

9. Which curve depicts the behavior of a heat-of-fusion salt?
   (a) 1   (c) 3
   (b) 2   (d) 4

10. Which curve most probably depicts the behavior expected for water storage?
   (a) 1   (c) 3
   (b) 2   (d) 4

11. The control system for a conventional, warm-water solar system is arranged to turn on the fluid in the solar collector when the temperature of the fluid leaving the collector is above (by a preset amount) the temperature of the water at the bottom of the storage tank. This is because:

(a) This assures a finite temperature drop across the collector.

(b) This maximizes the efficiency of the solar collector.

(c) This insures that the temperature of the fluid flowing into the storage tank will be higher than that of the water in the tank.

(d) This reduces the collector's upward heat loss.

12. For optimum storage of heat with rocks, small pebbles should be utilized because:

(a) This gives a large surface area to mass ratio.

(b) This gives a small surface area to mass ratio.

(c) This increases their specific heat.

(d) The economic cost of the rocks is decreased.

13. One of the following statements about using water as the heat-exchange fluid is not true.

(a) It is convenient to obtain and handle.

(b) It has a large heat capacity.

(c) A large amount of electrical power is required to run the water pump.

(d) Freeze protection can be obtained by mixing with ethylene glycol.

14. In a conventional, warm-water heating system, the auxiliary backup heating system should most certainly not be designed to do one of the following:

(a) Heat water in the delivery flow system.

(b) Heat air directly in the air delivery system.

(c) Heat water in the storage tank directly.

(d) Heat the circulating, solar collector heat-exchange fluid.

15. Eaves are recommended for south-facing windows because:

(a) They provide the necessary winter shade.

(b) They cut down on the solar input in the summer.

(c) They act as a reflector for winter optimization.

(d) They increase the solar input in the summer.

16. What percentage increase in light collection is possible in the winter with a reflector-collector system compared to that obtained with a normally tilted collector (with no reflector).

(a) 25%   (c) 95%

(b) 50%   (d) 200%

17. The optimum collector tilt angle when a horizontal reflector is used is:

(a) Near 0°   (c) Near 60°

(b) Near 45°   (d) Near 90°

18. Assume that a liquid-type, heat-exchange fluid is used in conjunction with a water storage tank. Describe three different ways that heat can be passed into the house using warm air as the heat transfer medium.

19. Describe three different arrangements for getting heat into the house from a water storage system using water as the heat transfer medium.

20. Suppose that you want to obtain approximately equal amounts of useful solar energy for space heating in the winter and air conditioning in the summer. Would the use of a reflector be a good idea? Explain.

21. Suppose that you wish to install a solar heating system into an existing home that has a sloping roof, with the sloping sides pointing east and west so that mounting a conventional collector on the roof would be impractical. The south wall of the house has no windows. Work up at least one design for a solar collector system in this geometrical situation.

22. Copper is the absorber material most resistant to corrosion. But even it can be susceptible to one type of corrosion. What kind of corrosion effect do you think this might be?

23. 8.9 cm of fiberglass insulation in a conventional 5- by 10-cm (2 in. × 4 in.) sidewall gives an insulation value of R-11. Also, 14 cm of fiberglass insulation in a conventional 5- by 15-cm (2 in. × 6 in.) sidewall gives an insulation value of R-19. What R value would be obtained if 5- by 20-cm studs were used in the sidewalls of a house?

24. Go to the library and look up the DD values in six different representative locations around the country. (e.g., look in the appendix of the book, *Energy, an Introduction to Physics,* by Romer). Calculate the minimum and maximum annual heating loads using the suggested range given in the text.

25. Describe three different ways that a backup auxiliary heating system can be combined with a regular solar system, having a water storage unit.

26. Design a preheating system for the building's hot water assuming that there is a water storage system.

27. Do the same as in the previous problem assuming the storage system consists of hot rocks.

28. Your solar collector is located in Medford, Oregon, and provides 40% of your annual heating load. How far away from the optimum can your collector be oriented so that you do not lose any more than 50% of the collectable solar energy?

29. Explain the advantages and disadvantages of controlling the operation of your solar collector by comparing the temperature at

the outlet of the collector with that at the top of the storage tank rather than at the bottom as described in the text.

30. Refer back to Problem 15, Chapter IV, on the economics of solar hot water heating. Using the knowledge gained in Chapters VII and VIII, make a quantitative estimate of the dollar cost per billion joules. Show how this is now a competitive number in any part of the country.

Active and Passive Solar Houses

Because of rapid growth in the use of the flat-plate collector solar system, one now has available a wide variety of approaches from which to choose. However, the basic principles involved, as outlined in the previous two chapters, remain the same. For this reason it is instructive to examine some of the solar houses built in the United States over the past 40 years. The experience gained in their construction and operation has been invaluable for the present development of the solar industry.

A.
EXAMPLES OF
ACTIVE SOLAR
HOUSES

From a functional standpoint, solar space heating systems are designed to operate in four basic modes. These are (1) direct space heating from the collector, (2) transfer of heat from the collector to the storage system, (3) space heating of the house from the storage system, and (4) space heating of the house from the auxiliary backup systems. In addition the solar system is often designed to preheat water from the incoming water supply prior to its passage through a conventional water heater. The domestic hot water, preheat system can be combined with the solar heating system or can be installed separately.

Solar systems can be classified into two general types: active or passive. Although there is some overlap between the two types, both classifications possess characteristics that are distinctly different from each other. An active system is one in which nonsolar energy is used to effect the transfer of energy. The collection, storage, and distribution of thermal energy is accomplished by moving various heat-exchange media with the aid of pumps and blowers. A passive system is one in which solar energy alone (for the most part) is used to effect the transfer of thermal energy. These two types of systems are compared in Figure 9-1, which shows a house being heated directly from the solar collector for an active and a passive solar system. The remainder of this chapter is devoted primarily to illustrative examples of these two basic types of solar systems.

A.1
MIT Solar Houses

In 1938 a Boston businessman, Godfrey L. Cabot, gave $600,000 to the Massachusetts Institute of Technology to be used for solar energy development. Over the next quarter century, MIT constructed four solar houses. The first solar house was constructed in 1939 under the direction of Dr. Hoyt Hottel, a prominent solar pioneer who also played a key role in later designs. The building was a one-story, two-room structure with a floor area of 46.5 m². The collector was about 38 m² in area, used water as the heat-exchange fluid, had triple glazing, and was tilted at only 30° to the horizontal. A 65,950-liter, water storage tank provided annual as well as diurnal storage. It is reported that the tank temperature reached 195°F at the end of the first summer. This solar house was demolished in 1941 to make way for other buildings.

The second solar structure was built in 1947 and was used only briefly for solar experiments. The building had seven south windows. Each one had its own separate collector (vertical orientation) and heat storage system. For example, one combination involved blackened cans located directly behind two cover plates.

This structure was converted in 1948 to the third MIT solar house experiment by enlarging the building and remodeling for occupancy. A 37-m² collector, again using water as the heat-exchange fluid, was mounted on the roof at an inclination angle of 57°. In the four coldest months this system provided about 75% of the space heating needs of

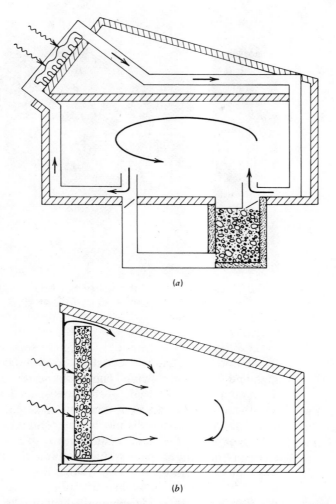

(a)

(b)

FIGURE 9-1   Solar radiation captured by the collectors and converted to thermal energy can be used to directly heat
the house. This process is illustrated for common
types of active and passive solar systems. (a) Active
solar system. (b) Passive solar system.

the 56-m² building. Electrical resistance heating was used in the 4550-
liter, water storage tank as the auxiliary energy source.

The last of the MIT solar houses was constructed in 1959 at Lexington, Massachusetts. The two-story structure, shown in Figure 9-2, had
a living space totalling 135 m². A portion of the first story was below

FIGURE 9-2   Picture of the fourth MIT solar house constructed at Lexington, Massachusetts.

ground to reduce heat losses. The 59-m² solar collector was tilted at 60° to the horizontal and was constructed from 60 panels, each 1.22-meters high and 0.81-meters wide. The solar thermal collector was a blackened aluminum sheet with two cover plates spaced 1.90 cm apart. Water was circulated through 0.95-cm copper tubes spaced 12.7-cm apart, which were fastened to the absorber plate. Heat storage was provided by a 5690-liter tank. The solar-heated hot water was pumped to a heat exchanger for heating air circulated about the house by a fan. The oil-fired, auxiliary water heater provided a reserve of 57 to 66°C hot water in a 1040-liter tank. Whenever the temperature in the large storage tank was too low to provide adequate heat to the house, the reserve tank was automatically called upon. The house was well designed to minimize heat losses, so that when the outside temperature was 0°F, only 32 million J/hr were required. Domestic hot water was also provided by the solar-heated reservoir whenever the temperature was adequate.

A thorough analysis of the system performance was made through the two winters of 1959 to 1961. Of the incident solar energy consisting of 311 billion J, about 32% was recovered by the solar thermal collector, and 29% contributed to useful heating of the house. This amounted to about one-half of the space heating needs of the house. Because of maintenance problems this system was abandoned after two years.

A.2
Solaris System

A number of pioneering solar homes were built in the Washington, D.C., area by H. F. Thomason. His first solar house was constructed in 1959 and had 112 m² of floor area. Only about 84 m² was actually heated in the winter. The solar collector had an area of 70 m², part of which was located on a 45° sloping roof, with the other portion located on a 60° sloping south wall. The absorber surface consisted of blackened, corrugated aluminum located below one layer of glass. About 5 cm of fiberglass was used as the insulating material on the back side. The heat-exchange fluid is water pumped from a 6000-liter storage tank by copper pipe to the top of the collector, where it is then allowed to trickle down the 2.5-cm-wide valleys of the black corrugated aluminum sheet. Maximum water temperatures achieved are in the 52 to 57°C range.

The water storage tank is a 1.22-meter-diameter, 5.2-meter-long steel tank, surrounded by 50 tons of egg-sized stones for additional thermal storage. The heat exchange medium for the house is air, which is heated by circulation around the tank and through the rock-filled chamber. This solar system provides about 85% of the building's heating needs. Thomason estimated a cost of $2500 for the entire system. Originally, summer cooling was obtained by removing heat from the rooms by circulating the house air through the previously cooled rock bin. The stored daytime heat was dissipated at night by circulating the warm water in the tank across a bare metal portion of the north roof. This nocturnal cooling proved inadequate for the Washington, D.C., area, and a 746-W refrigeration unit was installed to cool the tank water. This solar system is still in use.

A second solar house with a similar design was constructed by Thomason in 1960. This two-bedroom house has an area of 63 m² heated by 52 m² of solar collector. Both this building and the one described above were oriented 10° west of south. The trickling-water-type system used for the collector is identical to that for the first house except that the water-pump motor is only a 135-W unit, as compared to the 250-W motor used for the first solar house. In 1963 Thomason completed a larger, four-bedroom house with a swimming pool and game room. This house used 84 m² of collector, tilted at 60°, to heat a floor area of 139 m². Both the second and third Thomason houses use small reflectors to enhance the amount of sunlight entering the collector. The third solar house, has a reflector made of white-roll roofing spread over a 74-m² sun porch mounted above the swimming pool and in front of the solar collector. It was estimated that this solar system provides 75% of the space heating needs. Thomason has since built a number of solar homes and sells house plans through the Edmund Scientific Co. of Barrington, New Jersey.

A.3
Colorado Air
System

Dr. George Löf, another prominent solar pioneer, has played a leading role in the development of solar systems that use air as the heat-exchange medium. His first experimental house was constructed in Boulder, Colorado, in 1947. This building, with 93 m² of floor area, was partially heated by solar energy collected by 43 m² of solar panels sloping 27° upward from the horizontal. The solar heat, stored in 8.3 tons of rock, provided about 23% of the space heating needs.

His second solar house project was completed in 1958 and was financed by the American-Saint Gobain Corporation. The house is a nine-room residence comprising 190 m² of living space on the main floor. The house, illustrated in Figure 9-3, is of contemporary style. The solar collector consists of two 28-m² panels mounted at 45° to the horizontal. Recirculated air passes through the two sections in series, once upward and once downward. Each section of the collector is of the overlapped glass-plate design. Hot air at temperatures up to a maximum of about 79°C passes down to the basement where it first gives up some heat to the house hot-water supply, and then goes on either to the bottom of the storage unit or up through the furnace to the distribution system. The heat storage system consists of two cylinders of fiber board each filled with approximately 11 tons of small rocks. The rock column is 5.5-meters high, extending from the basement floor to the roof. Almost complete heat transfer from the air to the rocks is obtained from the large surface area of the rock in the storage chambers. The principal problems encountered initially with the design used in the Denver house were excessive glass breakage due to the defective system of glass-plate support for the solar collector, and excessive air leakage in the circulating system.

FIGURE 9-3 View of the solar home of G.O.G. Löf in Denver, Colorado. Air is used for both the collector heat-exchange fluid and the heat delivery system.

FIGURE 9-4   The University of Delaware solar house. Solar energy is collected with the aid of cadmium-sulfide solar cells. (Photograph courtesy of Dr. K. Boer.)

During the winter of 1959 to 1960, a detailed analysis of the daily heating performance was made. In a nine-month period, the total solar radiation on the collector surface was 240 billion J. Of this amount, 171 billion J was of sufficient intensity to warrant collection, and 59 billion Joules of useful heat was delivered. This represents an overall solar collection efficiency of about 25%. The 59 billion J represented 26% of the total of 280 billion J needed for space and water heating. While this is not a large fraction of the total heating requirement, it should be remembered that the ratio of collector area to floor area was also correspondingly small.

A.4
Solar One
Perhaps the most technologically advanced solar house is the one sponsored by the Institute of Energy Conversion at the University of Delaware and illustrated in Figure 9-4. Cadmium-sulfide solar cells are mounted on roof panels 1.3 m by 2.6 m. The solar cells convert part of the incoming solar energy directly into electricity. Behind these panels are channels for moving air to the basement. These channels serve the dual function of keeping the solar cells from overheating and at the same time allowing useful heat to be extracted and stored. The storage system consists of a system of plastic trays of eutectic salts developed by Maria Telkes, who has worked with heat-of-fusion storage over a long period of time. Electric energy from the solar cells is stored in

lead-acid auto batteries and used for lights and other purposes where the electrical energy can be utilized.

This program, directed by Dr. K. Boer, is aimed at coupling two different technologies—the generation of electricity directly from solar radiation using the photovoltaic effort and space heating using collected solar energy. The primary mission of the Institute is to develop cadmium-sulfide solar cells to the point where they can be mass produced cheaply enough for purchase by prospective home owners. The emphasis upon cadmium sulfide arises from the Institutes belief that they can be produced more cheaply than silicon solar cells. Disadvantages of cadmium sulfide cells include a low efficiency for the conversion of solar radiation to electricity (7% as compared to 11% or more for silicon solar cells) and deterioration of performance with time.

**A.5
Coos Bay Solar
House**

In 1967 Henry Mathew of Coos Bay, Oregon, designed and built a 139-m² solar house. The solar collector, shown in Figure 9-5, is distinguished by the use of a large reflector and an almost vertical collector. The reflector was used to enhance the intensity of the light entering the collector on days of low solar input that are often encountered in the Pacific Northwest. Mathew designed this house without any previous background in solar energy research.

Experiments were conducted in early 1965 on a 2.44- by 4.88-meter collector of the trickle type, with no reflector and a single plastic cover.

FIGURE 9-5   The Coos Bay, Oregon, solar home of Henry Mathew. This design is unique in that it uses a large reflector, sloped downward at 8°, in combination with an almost vertical solar collector.

It was installed next to an existing home, with the collector oriented at 70° to the horizontal and tied into an existing thermal storage system, using a 7600-liter water tank. This first collector provided adequate heat, but suffered from corrosion of the top feeder pipe because of the salt air environment existing at Coos Bay. Tests were then made on several other types of collectors. The crucial experimental criterion used was the time required for a drop of water to evaporate when placed on the hot collector plate. Final experimental tests were performed early in December 1965. A 75° tilted collector, combined with a horizontal reflector, was tested with one, two, and three cover plates; the collector with one cover reached the highest temperature. Measurements were made at noon on a cold, clear day. The absorbing plate was a piece of black steel, had 3.8 cm of fiberglass insulation behind it, and the first cover plate was spaced 1 cm above the absorber plate. Under the same weather conditions, the evaporation times were measured for the reflector-collector system with the collector tilted at 45°, 60°, 90°, and 110°. The fastest evaporation times were achieved with the collector in the vertical position.

Construction of the house and solar heating system was finished early in 1968. The house is one story, 27 × 11.3 meters, with the long dimension running east to west. Only the west half is lived in and heated; the east half is a combination garage, workshop, and greenhouse. The collector is 24-meters long and 1.52-meters high, tilted at 82° to the horizontal. The 30- × 6-meter roof in front of the collector was sloped at 8° and covered with aluminum foil pressed into hot roofing cement. The collector is tilted at 82°, rather than oriented vertically, to compensate for trees on low hills around the house, and to improve the appearance of the house by softening the angles to look more like a conventional roof. A second identical collector was installed a few years later on the ground near the house. It is estimated that about 80% of the annual space heating requirement is obtained with this combined system. A schematic arrangement of the Mathew solar system is shown in Figure 9-6.

B. PASSIVE SOLAR SYSTEMS

Another way to obtain solar space heating is to make the entire living space part of the collector and storage system. As noted above, systems that are characterized by a reliance upon natural convection and radiation for a major fraction of their heating requirements are known as passive. This does not rule out such devices as small fans or motorized insulation panels. But, the goal of a good passive solar design is to maximize the use of natural processes of conduction, convection, and radiation.

This type of approach is hardly a new idea. Some of the early Southwest native civilizations practiced such designs; Western ranch houses have often adopted important features such as a southward exposure with long overhangs for summer shade, low profiles, and heat-efficient

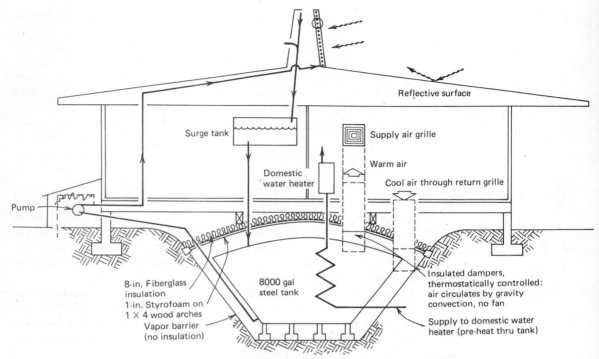

FIGURE 9-6   Schematic arrangement of the solar heating system designed and built by Henry Mathew of Coos Bay, Oregon. The solar collector is oriented at 82° to the plane of the horizontal, and the reflector is oriented 8° below the horizontal. Air circulation around the storage tank and through the house is due to the force of gravity only. Preheating of the hot water is obtained by passing it through the hot storage tank as shown.

local building materials. The renewed interest in solar energy has also spurred the development of passive systems.

With passive systems, the method of collection of solar energy is easy, since one just orients the house due south with a large window exposure. A calculation of solar gains through window areas is well understood. The exact amount of usable heat gained depends upon climatic conditions, the latitude, and the amount of insulation used. Properly oriented windows, combined with a means of insulation controlling heat losses during sunless hours, can often provide 20 to 50% of the annual space heating needs of a house. Shading devices must be used over the windows to exclude unwanted solar radiation.

The challenge confronting passive solar design is that of storage and control of the heat gained in this manner in order to maintain suitable comfort standards within the building. The approach customarily taken is to utilize a large storage mass. This is obtained through the use of dense interior materials with a high-heat-holding capacity such as masonry, adobe, concrete, stone, and water. These materials absorb surplus heat from the sunlight entering the windows and radiate it back into the room after dark. Some passive homes have walls up to a foot thick.

Designers are still grappling with the limitations of passive solar heating. Since direct sunlight cannot penetrate all parts of a typical house, an innovative architectural design is required. Some of these designs include having a shallow house on a long east-west axis, a ''stacked'' design like the Karen Terry house described below, or skylights that bring sunlight to interior spaces. Other innovations include movable shading devices to control sunlight, movable insulation panels to reduce nighttime heat losses, and ventilation ports to either augment or reduce daytime heating by means of natural convection.

Buildings that are designed to take advantage of the sun's energy can greatly reduce the demands on heating and cooling systems. This can be especially important in reducing the cost of the active solar heating and cooling system needed. Some of the important factors to be considered in passive solar design are as follows:

1. Buildings should face north-south rather than east-west to reduce annual energy consumption. South-facing windows receive maximum winter gains and minimum summer gains. East and west walls should not have large exposures of glass.

2. There are many different ways to eliminate heat loss through the windows at night or on cloudy days. For example, double-glazed windows or movable drapes can be used. An automatic method is to use the ''Beadwall'' system in which the windows are composed of two layers of glass separated by an air space of about 8 cm. When insulation is needed, light, plastic beads are blown automatically into the air space. The beads are removed by creating a small vacuum in the pipe system between the air space containing the beads and the adjacent storage bin. The use of some type of movable insulation can cut heat losses over that obtained with a simple window by as much as a factor of 5.

3. Since passive solar energy collection cannot store up heat as well as active systems, there is not as much carry-over for cloudy days. A large mass is needed to provide sensible heat storage. This is often

accomplished through the use of thick walls and floors. Another promising approach is through the use of a large surface area of rather thinner masonry. An outer layer of insulating material can be used advantageously.

4. Roof overhangs and window awnings can be designed to admit the low winter sun and block out the high summer sun. Deciduous trees and vines provide a natural seasonal control of shading.

5. Even during the winter months, the gain of heat close to large expanses of window area may be too large. One way to compensate for this is to use a fan to give forced circulation of the air. Another method is to orient the window area to the east of south. This reduces the late afternoon solar gain, at which time the house is already at a comfortable temperature. Another method is to use exposed masonry surfaces on the interior of the space, facing the windows. The masonry (e.g., brick or concrete) absorbs heat and can then deliver it to the house later on (see the following discussion of the Trombe wall design).

6. The use of specular reflectors can significantly increase the total influx of solar radiation through a window. Furthermore, a reflector oriented slightly below the horizontal plane will enhance the amount of light gathered in the winter while providing very little extra gain in the summer. Glare from a reflector below eye level can be a problem.

7. Glassed-in porches or greenhouses can be attached to a house in order to trap heat directly, exactly as in the case of windows. This arrangement permits a more exposed glazed area than for simple vertical windows. The greenhouse must be closed off at night to cut off heat losses when the sun is down.

The principles of passive solar heating are best illustrated by looking at a number of successful passive solar houses.

B.1
St. George's
School, Wallasey,
England

This building was built in 1962 as an annex to St. George's School in Liverpool, England at a latitude of 53° north. It was designed by A. E. Morgan to house 300 students and to gain its entire heating needs from solar energy, lighting, and the students. A schematic drawing of this structure is shown in Figure 9-7. The basic design approach was to make the structure sufficiently massive so that thermal inertia would prevent rapid or extreme fluctuations in the internal air temperature. In addition, the exterior walls are insulated with 13.5 cm of polystyrene.

The building is a long, thin two-story structure running east and west. The construction is concrete, with the ground floor consisting of 10 cm

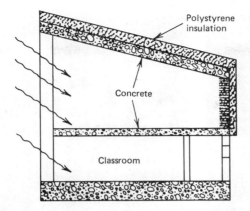

Polystyrene insulation

Concrete

Classroom

FIGURE 9-7   Cross-sectional view of the Wallasey annex to St. George's school. Sunlight enters through the large double-glazed south wall. Stable temperatures are maintained through the use of massive floors and walls and by the use of polystyrene insulation to help reduce building heat losses.

of screen upon 15 cm of concrete, the intermediate floor is a 22-cm slab of concrete, and the roof is 17 cm of thick concrete. The partition walls are 22 cm of plastered brickwork. On the north side the external walls are also 22 cm of brick, with an additional 12.5 cm of polystyrene insulation. At the ground floor level, the external wall, shown in Figure 9-8, is mostly double-glazed and is 8.2-meters tall and about 70-meters long. The outside sheet of glass is clear and is spaced about 0.61-meters away from the inside sheet of "figured" glass—a type of glass that refracts the sun's rays in such a way that both the ceiling and the floor are irradiated in a uniform manner. Each classroom, however, is provided with two or three openable "view" windows, which constitute areas of single glazing. Measurements of the thermal performance of the system indicate that the maximum fluctuations during winter days is about 2 to 3°C.

**B.2
Karen Terry's Sante
Fe House**

One way to overcome the problem arising because direct sunlight cannot penetrate all parts of a typical house is to use a "stacked" design. This approach was taken by architect David Wright in designing the Sante Fe solar house of Karen Terry shown in Figure 9-9. The house is oriented to receive the low-winter sun, receiving and storing heat on all levels, as indicated in the sectional view shown in Figure 9-10.

The solar house is set deep into the hillside with earth berms piled up against the side and the back. The house measures 5.5 by 16 meters with 37 m² of double-glazed glass in four levels. During the summer the glass is shaded by removable wooden louvers. Between the lower and

FIGURE 9-8   Photograph of the Wallasey annex, showing the large south-facing expanse of glass.

FIGURE 9-9   Picture of the passive solar home of Karen Terry in Sante Fe, New Mexico.

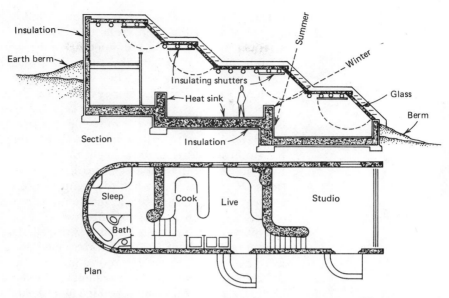

FIGURE 9-10   Plan and sectional views of the Terry solar house. The living space is arranged on three levels, while the fenestration is on four levels.

middle levels is a wall consisting of eighteen 208-liter steel drums containing water and an anticorrosive additive. Similar drums are used to form the barrier between the middle and upper levels. At night, 5-cm polystyrene insulating shutters swing up to cover the glass, helping to hold in the heat that has been absorbed during the day. The normal temperature range on a typical winter day is from 15°C at sunrise to 24°C in the late afternoon. Backup heat is provided by a fireplace and a woodstove.

B.3
Roof Ponds
Another simple and elegant passive solar system uses roof ponds. In this system water is contained in a clear plastic bag and placed on a black metal roof. This idea was designed and patented by Harold Hay who first got the idea while in India on a technical aid mission for the U.S. government. He noticed that many people lived in rusty, sheet-metal shacks, which were hot in the day and cold at night. Hay's innovation was to remove the insulation from the roof on winter days so that the roof would get hot. Replacing the insulation at night would trap this heat, allowing it to provide useful heat throughout the night. During the summer a reverse procedure would be followed, letting the house cool at night and replacing the insulation in the daytime to keep out the heat.

Over the years this idea was refined and in 1966 and 1967, Hay, in collaboration with John Yellott of the University of Arizona, constructed a 3- by 3.7-meter building using water basins as the actual roofing material. During the summer, a slab of foam insulation was rolled back at night, and the water became cold through night sky evaporation. Since the water reservoir sat directly on a metal ceiling it absorbed heat from the room and kept the building air conditioned all day. During the winter the movable insulation was rolled back during the day in order to collect heat. It then radiated enough heat into the house through the ceiling at night to keep the room comfortable. Tests over an 18-month period seem to have justified the designer's predictions. Hay notes that when the temperature outside was 36°C, he had to wear a sweater and a coat inside to keep warm.

This approach should be effective in dry climatic regions. An improved version has been tested in Atascadero, California. The exposed roof area was about 102 m², approximately equal to the floor area. The average water depth of the roof ponds was about 20 cm, giving a total of 22,700 liters of water. The system was operated both with and without glazing. The glazing, an inflatable clear plastic cover, proved necessary in order to maintain the indoor temperature during the winter months. This solar system supplied 100% of the house's heating and cooling requirements. The indoor temperature was maintained between 14 and 23°C. The weight of the roof pond apparently created no major problems. Problems with roof leaks developed, but were corrected.

B.4
The Trombe Wall

Another novel passive solar approach uses a large, massive wall for thermal storage. This is exemplified by the Trombe-wall houses constructed in the solar community near Odeillo in the French Pyrenees by Felix Trombe and his collaborators. The design principle is illustrated in Figure 9-11 and features a massive south-facing, vertical concrete wall, which is painted black and covered with a sheet of glass. An airspace runs between the concrete and the glass; the chimney effect causes the heated air to rise. During the day openings at the bottom and top of the wall allow the cold air along the floor to enter the air space. Similar openings at the top provide an opportunity for the warm air to reenter the room behind the wall. The air then circulates through the room, warming the walls and the occupants with a portion of the energy collected.

The remaining energy is transmitted by conduction through the wall, resulting in a rise in temperature at the wall's interior surface. Energy is then spread through the living space by conduction, convection, and radiation.

Data taken from a Trombe-wall building constructed in 1967, which had walls 0.60-meter thick, indicated that one-third of the solar radiation incident on the south wall during the winter months was transferred into the house. About 70% of the space heating load was provided by

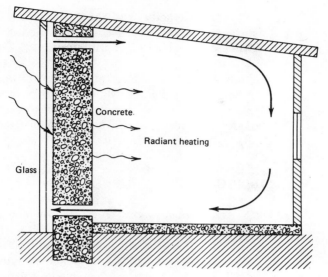

FIGURE 9-11   Trombe wall-type of passive solar house. The house faces south with the air heated by the incoming solar radiation, which is absorbed by the black face of the wall.

solar energy. Of this amount, about 20% was transported into the living space by convection, and the remaining 50% came by conduction through the wall. In the summer, the glazed concrete wall was shaded from the sun by the overhanging roof.

A detailed analytical study of Trombe walls has been made by J. D. Balcomb and colleagues at the Los Alamos Scientific Laboratory to determine the effect upon performance of various parameter changes. The validity of the computer simulation studies was established by comparing the predicted results with experimental data from several small experimental passive test rooms. The general conclusions reached were as follows:

1. For the same glass area and thermal-storage heat capacity, a water wall will slightly outperform a masonry wall.

2. For best results, either double-glazing or night insulation (or both) is needed.

3. Best performance with a masonry Trombe wall is obtained when the thickness is 20 to 50 cm. With smaller thicknesses the wall has too

little heat capacity, and the temperature fluctuations on the inner surface are excessive. No such restriction applies if the wall is constructed of water-filled barrels or tanks.

4. Vents at the top and bottom of the wall to allow natural circulation of air increase the overall annual performance of the wall. Nevertheless, the vents must have provision to prevent reverse flow, or the overall performance will be less than if the vents were not used at all.

BIBLIOGRAPHY

1. *Sunset Homeowner's Guide to Solar Heating* (Menlo Park, Ca.: Lane Publishing Co., 1978).

   Contains an excellent and readable discussion of many features of solar heating systems. This is a very nicely illustrated book.

2. G. O. G. Löf, "The Heating and Cooling of Buildings with Solar Energy," Chapter 11, in *Introduction to the Utilization of Solar Energy,* edited by A. M. Zarem, and D. D. Erway (New York: McGraw-Hill Book Co., Inc., 1963).

   Contains a useful description of some of the historically important active solar houses.

3. W. Shurcliff, *Solar Heating Buildings, A Brief Survey,* (Cambridge, Mass., 1976).

   Lists a large number of solar heated buildings in the United States and abroad. Considerable technical information, along with a sketch, are given for each building.

PROBLEMS

1. Circulation of air from the house to the storage tank in the H. Mathew solar house is accomplished by means of:
   (a) Gravity.
   (b) A fan.
   (c) Combined result of using a small fan and hot air ducts that are placed lower than the cold air ones.
   (d) A water pump plus heat exchange in the house.
2. One of the following is not a key element in the design of a passive solar house.
   (a) Large direct solar gain.
   (b) Roof sloped at a tilt angle of latitude plus 10°.

(c) Large mass to provide for thermal heat storage.

(d) Good insulation to minimize heat losses.

3. Discuss the possible weaknesses in the first MIT solar house. Explain carefully your reasons.

4. What is the principle advantage of the "trickle-type" collector of H. Thomason? Can you think of any major weaknesses in the Thomason collector design?

5. Explain the principle involved in using nocturnal cooling in the summer. This approach was utilized by H. Thomason in Washington, D.C. However, the amount of cooling obtained there in the summertime was insufficient. How do you account for this? What type of climate is best suited for this method of cooling a house?

6. Do you think liquid heat-exchange systems are the best type for cloudy climates? Explain. Contrast with air systems.

7. What is presently the major obstacle to widespread utilization of solar cells for space heating?

8. Henry Mathew performed his experimental evaluation of various collector configurations for the Coos Bay solar house by measuring the time required for a drop of water to evaporate when placed on the hot collector plate. Explain how this test works.

9. Devise an alternative method to that outlined in Problem 8 for the experimental comparison of various reflector-collector geometries. Assume that the only measuring equipment available to you is a thermometer.

10. Why is a large mass necessary for a good passive solar house?

11. What are the three most important design features for passive solar heating?

12. What is the purpose of the polystyrene on the Wallasey solar building?

13. Explain why the performance of a Trombe wall, constructed of water-filled tanks, does not decrease as the wall thickness increases beyond 50 cm, in sharp contrast with the performance of a masonry wall, which becomes less effective as the wall thickness increases beyond 50 cm.

Solar
Thermal
X Power

The Energy Research and Development Administration (now in the Department of Energy) has prepared a comprehensive plan for energy research to provide for the nation's energy requirements. This plan, as presented in the document ERDA-49, projects that a total of 45 quads (1 quad $\cong 10^{15}$ Btu, or about 10^{18} J) will be supplied by solar energy by the year 2020. This same projection calls for 135 quads of energy from coal and nuclear power, requiring perhaps six hundred 1000-MW nuclear reactors and the annual consumption of 3 to 4 billion tons of coal. This large estimate for the U.S. energy consumption can be questioned. But it is apparent that solar electricity is now being counted upon to provide for an appreciable fraction of our energy use by the year 2000. This is a remarkable turnabout from the prevailing sentiment of 10 years ago that using the sun's energy was only a romantic dream of those who preferred to return to a more primitive existence.

As we have just seen in the past few chapters, utilization of the energy coming to us from the sun for the purposes of space heating and cooling is well underway. This is due to the solid technical base provided by the results of flat-plate collector and solar house experiments during the past 35 years. Future developmental work will be aimed at technical improvements such as freeze protection, selective surfaces, reducing the economic cost of solar components, and improving their long-term efficiency and reliability. Considerable research effort also needs to be done on such topics as specialized approaches for specific climatic regions, the possible use of heat pumps, passive solar houses, and heat-actuated refrigeration devices. Nevertheless, the basic technology for the rapid development of the solar heating and cooling industry is available and is presently either already competitive with conventional approaches or on the verge of being competitive in most areas of the country.

No such sound technological base exists for the larger-scale development of solar power and electricity. Many different small, solar engines have been constructed over the past 100 years. But, a very large amount of technical effort remains to be done in order to develop economical solar electricity. The different approaches to the generation of power from solar energy may be roughly divided into the following categories:

1. *The direct conversion of solar radiation into electricity using solar cells.* Photovoltaic devices are presently economically viable for remote site applications. Their widespread use will depend upon the results of the present massive federal research and development program. This topic is of sufficient importance that the next chapter is devoted entirely to the subject.

2. *The conversion of thermal energy into electricity by a large number of distributed collectors employing a moderate degree of concentration.* This approach enhances the incident solar intensity roughly by a factor of 10 to 100 when reflectors are used, making it possible to obtain temperatures over the range of 100 to 500°C. Intermediate-temperature systems have the advantage that they can be fairly small in size and can be located close to the ultimate energy user.

3. *The central receiver concept.* A heat-exchange fluid located on top of a tall tower is heated by the solar energy reflected from thousands of individual mirrors that track the sun (heliostats). This approach would involve concentrating the incident solar intensity by a factor greater than 1000. The resulting elevated fluid temperatures permit a high-thermodynamic efficiency for the conversion of thermal energy to useful power.

4. *Generation of electricity from biomass.* Photosynthesis is nature's way of converting solar radiation to a useful form. Many different schemes for extracting the energy from biomass are presently under study. This topic is not discussed in this text.

5. *Ocean thermal energy conversion.* In this approach power will be generated in a turbine, driven by the small temperature gradients

present in the ocean between the surface and the water at depths of hundreds of meters.

Before going on to discuss some of these schemes in more detail it is worthwhile to note a few of the major problems associated with the large-scale generation of power using solar energy. Since the incident energy density is low, solar energy power plants must collect light over a large amount of area, thereby requiring large initial expenditures for equipment. The capital expense for the collecting apparatus and power conversion devices is the dominant feature in determining the ultimate cost per kilowatt-hour of electricity, since the fuel is free. The only real pollution problem to be faced is that of the disposal of waste heat; solar energy is no different from other sources of energy in this respect. On the other hand there is no emission of particulate matter or noxious gases to contend with and no radioactive waste to worry about. Solar power plants will be located at large distances from the major population centers, so that transmission losses and costs must be carefully considered. Perhaps the worst feature associated with large-scale power generation from solar energy has to do with the daily collection period. Solar energy will be collected only during a few daylight hours. Either the power generated must be used to augment conventional sources, or adequate means of storage of energy must be developed.

The solar power schemes to be discussed here all involve the conversion of heat energy to electricity. This conversion process requires an engine working in a cyclic fashion. To do this some working material, say hot water or steam, must undergo a series of changes of pressure, volume, and temperature. When this happens we say that a thermodynamic process has taken place. The entire subject is complicated because of the fact that one is not dealing with a simple system of a few molecules, where the laws of mechanics can be applied. Instead, one is dealing with incredibly large systems of particles so that only the average behavior of such a system can be described. Despite the complexity, the description of such systems was fully understood by the end of the nineteenth century. Practical thermodynamics, from an engineering standpoint, can be quite complex and detailed. For our purpose, which is to understand the basic requirements for generating electricity, comprehension of a few fundamental concepts will be sufficient.

**A.
THERMO-
DYNAMICS**

In Chapter II it was pointed out that the usefulness of the concept of energy is due to the fact that the total energy of a system is conserved. When energy is used to perform useful work, what occurs is that energy is converted from one form to another. When gasoline is burned in a car, the stored potential energy (chemical) in the fuel is converted into useful work—the motional energy of the auto. But, this example raises a serious question. A quantitative measurement of the useful work done upon the car by burning the gasoline will show that only a fraction of the stored energy in the gasoline has been converted to useful purposes. The remainder has been expended in friction or has been exhausted into the atmosphere. This seems like a shameful waste. Can human beings do something about this wasteful use of energy?

In this connection it is useful to review the efficiency of various ways of generating electricity. Remember that the efficiency of a device is defined as the ratio of useful work output to the total energy put into it. The efficiency of a typical coal-fired, electric power generating plant is typically about 40%. This means that 60% of the available, stored energy in the coal has been rejected to the outside world. Nuclear reactors commonly have efficiencies in the 32% range, giving an even larger fraction of energy rejected. On the other hand, when water passes over a dam and is used to generate electricity, it is found that almost 90% of the gravitational potential energy of the water is converted to electrical energy. This is a tremendous improvement over the other ways of generating electrical power and leads us to ask exactly what is responsible for such wide variations in the efficiency of power generation.

One possible explanation might be that there has been a breakdown in the law of conservation of energy; however, since there is no evidence for this in any other measurement in physics, we rule out this possibility. Furthermore, we also rule out friction losses in the turbines, etc. Good design can often reduce the friction losses to a few percent, so that this cannot be the reason for the relatively low efficiency of coal and nuclear power plants as compared to hydroelectric power. Instead, the solution to this puzzle lies at a more fundamental level. The low efficiencies are due to a fundamental thermodynamic limit on the efficiency of conversion of energy to useful work by any process involving a transfer of heat. This limit constrains the possible efficiency of conversion of heat to work, such as in the generation of electricity. However, there is no such constraint on the reverse process, since work can be converted to heat with an efficiency of 100%.

The detailed explanation of this process involves some very subtle ideas. Before beginning, it is useful to first discuss the concept of conservation of energy carefully in the context of processes involving heat. As noted briefly in Chapter II, until the middle of the nineteenth century, it was thought that heat was somehow proportional to the amount of a fluid (called caloric) that was contained in a substance. This caloric theory of heat was able to explain many phenomena. For example, if more caloric flowed into a substance then it should become warmer, exactly as observed.

However, there were problems. If heat is a fluid and is squeezed out of a piece of material, there should be less heat left in the material. This results in the material having less capacity to hold heat. Yet no such effect was ever found. The first scientist to quantitatively show that heat was just another form of energy was Benjamin Thompson (Count Rumford). He became interested in the question of what exactly is heat while supervising the boring of cannons. This was performed by a number of oxen, who traversed a circular motion about the cannon in order to turn the mechanical device responsible for boring the cannon. By careful computation (for these circumstances) he found that the heat generated in the cannon, which was obtained from the rise of temperature, was proportional to the mechanical work done by the oxen in a fixed time period—no matter how many hours or days the process had been going on. He concluded that heat is associated with microscopic motion. According to the caloric theory, the chips bored from the brass cannon barrel would be hotter than the barrel because the bore would squeeze the caloric out of the barrel along with the brass chips; in fact, Count Rumford discovered that both the bored barrel and the chips were at the same temperature. Several years after the experiments of Count Rumford, Joule made a careful series of measurements in order to determine quantitatively the relation between heat energy in calories and mechanical energy in joules. The present-day accepted equivalence is: 1 cal = 4.184 J.

A.1
The First Law

It is now understood that heat is a form of energy. More precisely, it is the transfer of energy by molecular collisions. Scientists call the energy content of a substance its internal energy and save the word heat for the gain or loss of internal energy brought about by molecular collisions. For example, if a hot iron poker is dipped into cold water (Figure 10-1), the iron molecules lose energy while the water molecules gain energy. Energy has been transferred from the hot poker to the water molecules as a result of a large number of molecular collisions. The internal energy of a substance is proportional to the energy of the random motion of the molecules of the material. In turn, this motional energy is proportional to the parameter we have called temperature. In other words, the energy content of a substance is proportional to its absolute temperature (or vice versa).

The statement that energy is conserved is an empirical law of physics. When this law is extended to include heat and internal energy, it is called the first law of thermodynamics. We demonstrate the use of the first law with the aid of the simple system shown in Figure 10-2. Suppose that a quantity of heat is added to this system. For example, this could be done by using the flame of a bunsen burner. This additional heat will warm up the gas causing its molecules to move faster. In other words, the internal energy of the gas will have increased as measured by its increased temperature. At the same time, the increased thermal motion of the gas molecules cause the pressure on the piston to increase. If the

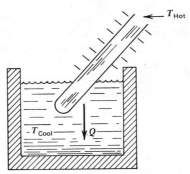

FIGURE 10-1   The energy transferred by molecular collisions between the hot poker and the cooler water is called
heat. The energy content (due to random molecular
motion) of the water or the hot poker is termed the
internal energy of the substance. The flow of heat,
Q, is from T_{HOT} to T_{COOL}. At equilibrium the temperature of the water and the poker will be identical. This
resultant equilibrium temperature will be intermediate between the two initial temperatures.

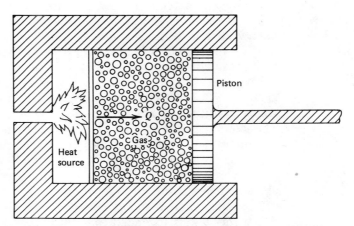

FIGURE 10-2   Apparatus demonstrating the first law of thermodynamics. If the piston is held fixed no work is done,
and Q equals the change in internal energy of the
gas. However, if the piston is free to move, then some
work is done and conservation of energy tells us that
the resultant increase in internal energy of the gas
is smaller than for the case where the piston was
fixed.

piston is held fixed, no work will be done and all of the heat energy ΔQ will go into increasing the internal energy of the gas, resulting in a temperature increase from its initial value to a final value T_1. On the other hand, if the piston is free to move, a certain amount of work ΔW will be done in displacing it upward against the force of gravity. The heat energy ΔQ will then be divided in some manner between the work ΔW and the change in internal energy. In this case, the first law, which expresses conservation of energy, tells us that the increase in internal energy will be smaller than in the first case, so that the final temperature of the gas, T_2, will be smaller than T_1.

A.2
The Second Law

Let us now return to the question about the efficiency of generating stations that was raised at the beginning of this section on thermodynamics. Consider a specific example of one type of heat engine used for large-scale electric power generation, the steam turbine. The basic idea can be understood with the aid of Figure 10-3. Steam is heated to a condition of high temperature and pressure and allowed to expand against the blades of the turbine. This expansion causes the blades to rotate, so that heat energy has been converted to useful work, in a manner analogous to the linear motion of the piston (Figure 10-2), under the influence of the expanding gases. The source of heat could come from nuclear reactions, the burning of oil or coal, hot water or steam

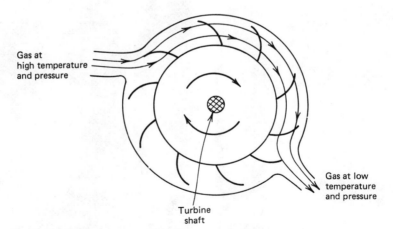

FIGURE 10-3   Example of a steam turbine used in thermal power plants to produce electricity. The hot gas expands against the turbine blades to produce rotation, which in turn leads to electric power. A typical efficiency for a coal-fired power plant is 40%.

from a geothermal source, or from some type of solar energy collector. In the case of the turbine not all of the energy contained in the hot steam is converted to useful work (the rotating blades). The outlet gas that is at a finite temperature has a quite significant amount of internal energy. Consequently, typically only 30 to 40% of the incoming energy is actually converted to useful work. This performance is to be compared with the efficiency of a hydroelectric turbine such as that shown in Figure 10-4, which is almost 90%.

Why is the turbine efficiency so great in the hydroelectric case? The answer is found in the fact that the motion of the falling water is orderly, with the water molecules having a dominant velocity component along the stream flow, so that all of the water falls in a given direction. This is not true for the internal energy of molecules in a hot gas where the motion is random, or disordered. The results of many investigations of thermodynamic processes have shown that nature has a general rule stating that there is a tendency toward disorder, provided that one deals with large enough samples of matter where more than a few molecules are involved. Thus, the rule is

$$Order \rightarrow Disorder$$

Because no exceptions to this tendency have been found (for a closed,

FIGURE 10-4   Turbine in a hydroelectric power plant. The overall efficiency for converting the stored potential energy of the water to electric power is of the order of 90%.

complete system), it is accepted as a law of nature and is called the second law of thermodynamics.

Since mechanical energy of motion is more ordered than is thermal energy, it can always be completely converted into other forms, resulting in the very high efficiency found in the hydroelectric turbine.

On the other hand, disordered thermal motion cannot be completely converted into ordered mechanical motion. In the case of the conversion of thermal energy in the steam turbine of Figure 10-3, only a fraction can go to ordered mechanical energy while the remainder must be in the form of even more disordered thermal motion of the ejected exhaust gases. In other words, only a portion of the input energy in a thermal electric generating plant can be converted to useful work (in the form of electric power), and the efficiency of such plants is much lower than for hydroelectric generation.

The above description of the second law of thermodynamics in terms of ordered and disordered motion is rather abstract. Thanks to the work of a brilliant young French field artillery officer and physicist, Sadi Carnot, the second law can be restated in an entirely different manner. To do this it is necessary to introduce the concept of a thermodynamic cycle. For example, the fluid used in a steam turbine undergoes a cyclic process in which the steam starts out in some "state" (by state we mean some condition of pressure, volume, and temperature), undergoes a thermal process whereby work is done, and then finally is returned to the initial state. In a steam engine, water absorbs heat in the engine's boiler during the process of evaporation to steam. The steam then expands and performs mechanical work. After the work has been performed, the steam is condensed back to water as heat is removed from it in the condenser. Then the cycle is repeated as the water is pumped back into the boiler and converted to steam. This cyclic process is illustrated in Figure 10-5. The importance of going through a cyclic process is that no net change of internal energy of the fluid (or gas) occurs. In other words, the net heat transferred is equal to the work done by the gas.

Since the change in internal energy is zero, we have

$$(10\text{-}1) \qquad\qquad W = Q_1 - Q_2$$

which just expresses quantitatively that the work done equals the difference between the heat input to the fluid and the heat delivered to the cooler output. The efficiency of the engine is then

$$(10\text{-}2) \qquad\qquad \eta = \frac{\text{work output}}{\text{heat input}} = \frac{W}{Q_1} = \frac{Q_1 - Q_2}{Q_1}$$

Carnot worked out the relationship between heat and temperature for

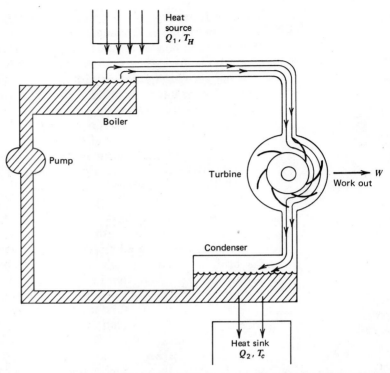

FIGURE 10-5 Illustrates how the fluid in a steam turbine undergoes a cyclic process, returning finally to its initial state. The useful work out is thus equal to the difference between Q_1 and Q_2 (assuming no losses).

an ideal engine—one with no friction losses—so that the heat level at each end of the cycle would be equal except for the heat energy extracted to perform mechanical work. This concept is important because any real engine has a lower efficiency than the ideal one. Carnot showed that for an ideal engine,

(10-3)
$$\frac{Q_1}{Q_2} = \frac{T_1}{T_2}$$

If we substitute this relation in the expression above for the efficiency, we have

(10-4)
$$\eta_{\text{ideal}} = 1 - \frac{T_2}{T_1}$$

For an efficient engine the heat source needs to be as hot as possible, and the exhaust temperature should be as low as possible.

The second law of thermodynamics can be stated in many equivalent ways. We have presented it above in the form of a statement about increasing disorder. The essential content of this law is that there is something about the real world that prevents us from extracting heat from a reservoir, converting it to work, and then cooling the reservoir. One of the ways in which the second law can be stated is as follows:

It is impossible to construct an engine, operating in some continuous fashion, which does nothing other than take heat from a source and perform an equivalent amount of work.

B. SOLAR TOWERS

Because of problems associated with integrating a solar electric plant into the electric power grid, it appears that a plant size of 30 to 100 MW is the lower limit to be considered. One way to do this is to use a system of distributed collectors (see Section C). The energy absorbed by the distributed collectors can then be transported to a common point where electric power could be generated with a single turbine, or a series of small turbines could be used with each small, distributed collection field. But, it is now commonly believed that for larger plants these approaches will be less efficient and less economical than the approach of using a single solar tower.

In the solar tower scheme a central receiver will be placed at the top of a tall tower located near the center of a field of tracking mirrors (heliostats). The reflected energy is absorbed by the receiver and used to heat some type of fluid or gas. The high-temperature, high-pressure fluid is returned to the ground and used to power an electric turbine generator and simultaneously to deliver energy to a thermal storage unit for deferred operation. The mirror field for a 100-MW facility is presently envisaged to consist of about 20,000 heliostats, each with an area of 40 m², deployed over an area of rougly 3.5 km². The tower for a power plant of this size would be 260-meters high.

In order to achieve good efficiencies for a thermal power plant, it is necessary to have the source of heat at medium to high temperatures. Since the incident solar radiation has a low energy density, some means of concentrating the incoming solar radiation to a much higher energy density must be devised. The diffuse radiation comes from all directions in the sky, so that only the direct solar beam can be concentrated; solar electric plants will naturally be located in areas such as the Southwest where the direct solar radiation is large. As noted above, the solar tower concept achieves a high concentration factor through the use of a large field of reflecting mirrors. In this case the solar energy density upon the central receiver will be over 1000 times as great as that for the normal incident solar radiation. The temperature achieved at the focus of the mirrors (the receiver) will be of the order of 800 to 1200°C. It is quite difficult to increase the temperature markedly above these values. Increasing the concentration factor from 1000 to 10,000 would only increase the temperature of a system, initially at 1000°C, to about 2000°C.

The development of large solar furnaces in France under the direction of Felix Trombe was a precursor of the present power tower effort. At the end of World War II, Dr. Trombe interested the French government in building a solar furnace for research at Mt. Louis in the Pyrenees. This was an area with clear air and large annual solar flux. Trombe, who began his research in this area in 1946, was convinced that solar furnaces had enormous possibilities for high-temperature research. His first research equipment consisted of converted German antiaircraft searchlights and parabolic sound collectors, used in wartime for detection of enemy aircraft engines.

The Mt. Louis solar furnace was completed in 1962. In this device, the sun's radiation was reflected by a large, flat mirror that tracked the sun. This large mirror consisted of 168 separate plane mirrors of silvered glass, each 50 by 50 cm in size. The solar radiation was reflected from this large mirror onto a second concave reflector 11 m in diameter, made up of 3500 small reflectors, and located 21 m from the first reflector. Finally, the solar energy was focused by the parabolic reflector onto a point where 50 kW of heat were developed and used both for research and as an industrial smelter. Later, Trombe constructed an even larger, 1-MW (thermal power) solar furnace at Odeillo in southern France. In this device, the sun's rays are reflected by 63 heliostats, covering an area of 2835 m², and located on the side of a hill. The reflected radiation is then directed upon a parabolic concentrator designed as an integral part of a nine-story building. The parabolic reflector, shown in Figure 10-6, is 40 meters high and 54 meters across. The peak flux at the focus of the parabolic concentrator is 16 million W/m² or about 17,000 times the standard clear-day insolation of 950 W/m². For many years this instrument was used for a wide variety of high-temperature experiments. In 1973, the transformation of the solar furnace into an electrical power plant was begun; in November of 1976, solar electricity flowed into France's electric grid. Since then, the facility has been reconverted to one for research at high temperatures.

In Italy, Francia developed an intricate clock-driven field of 271 heliostats and was able to produce steam at a rate equivalent to 150 kW. This concept is not thought to be suitable for large solar electric plants because of the difficulties associated with connecting many thousands of heliostats into a single, clockwork device. A 400-kW (thermal

FIGURE 10-6   Picture of the 1-MW solar furnace constructed at Odeillo, France, under the direction of F. Trombe. The north side of the building has been formed into the shape of a large parabolic reflector.

power) test facility of this type has been constructed at Georgia Technological University. This solar thermal test facility has 550 heliostats, covering an area of 532 m², resulting in a peak solar energy density that is 4000 times the standard value (950 W/m²).

Because the heliostat field comprises almost 50% of the total cost of a solar tower system, good design is crucial. A typical system might consist of a small heliostat mounted on a pedestal and positioned by an individual automatic control system. Each reflecting mirror then images the sun onto a fixed surface. The pedestal supporting the mirrors would have two degrees of freedom of motion providing rotation and tilt. Each morning the rotation and tilt positions on the mirrors would be indexed to their proper position, and the automatic control system would then maintain the reflected energy on the receiver located at the top of the tower. For aerospace companies, who are used to designing highly accurate tracking mechanisms for ballistic missiles, the degree of accuracy required here can be easily and inexpensively attained.

The heliostats of a solar tower will be constructed to meet exacting tolerances. Four different heliostat designs, shown in Figure 10-7, are presently being tested. All except the Boeing design have steel and glass construction. The Boeing mirror is made of aluminized polyester stretched across a circular frame. Boeing believes that the 20% transmission loss that occurs when sunlight passes through the plastic bubble enclosing the mirror will be more than compensated for by the cost reduction achieved through the use of lightweight materials. The Honeywell design uses rectangular mirrors mounted on a geared tracking

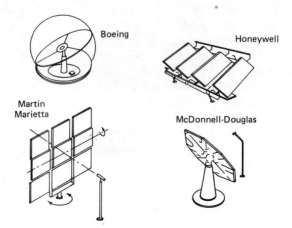

FIGURE 10-7   The heliostat designs proposed for the 10-MW solar tower pilot plant to be built near Barstow, California.

frame that tilts in two directions. The heliostat of McDonnell-Douglas is a solid dish formed of eight mirror segments. The entire assembly is mounted on a radar pedestal. Martin-Marietta uses nine mirrors mounted on a common tracking frame.

The heliostats must be spaced in such a way as to avoid shading and blocking of the reflected radiation. The optimum configuration is obtained with a nonuniform mirror distribution in which about 25% of the land area encompassed by the mirror field is actually covered by mirrors. The density of the reflecting mirrors decreases with distance from the solar tower in order to prevent blocking of the reflected sunlight by adjacent heliostats. Present (1977) costs for the heliostats are quite high, lying in the $500 to $1000 per square meter range. Since the heliostats represent such a large fraction of the total cost, it is imperative to drastically reduce their cost. Through improved techniques and mass production economies, the Department of Energy hopes to reduce the heliostat cost to $70 to $80 per square meter.

The toughest technical problem to be overcome may be that of the central receiver. Because of the high-concentration factor, the power density inside the receiver will be quite high and variable. Some designs affect slightly the focus of different groups of heliostats in order to spread out the power profile. The materials used in the receiver must be able to withstand instantaneous changes in energy densities from 0 to 5 MW/m^2. Because of the diurnal variation of the incoming radiation, the system will be subject to frequent temperature cycles. One design concept under study would use a gas, rather than a steam turbine. In this case, the best operating temperature is twice that of a steam turbine. This would rule out the use of metals in the high-temperature face of the receiver; the use of ceramic heat-exchange tubes has been suggested. The use of a dry gas turbine has the strong advantage of reducing the environmental impact as compared to wet systems, which will require the importation of water to the site.

The first major step in the federal research and development program to implement commercial solar electric power using the tower approach involved the completion and testing of a 5-MW (thermal) facility located at Albuquerque, New Mexico. A mirror field of 222 heliostats covering 8257 m^2 was utilized. A peak flux of 2.5 million W/m^2 has been obtained, representing a concentration factor of 2630 over the standard value. Figure 10-8 shows a view of the partially completed facility. The tower is 61-meters tall and was constructed from 4400 m^3 of concrete. Each heliostat is anchored with nine 10-ton concrete footings, and the entire mirror field covers 40 hectares. The next stage in the federal development program is the construction of a $130 million, 10-MW electric test facility at Barstow, California. The tower for this project will be about 152-meters high.

FIGURE 10-8   View of the partially completed 5-MW (thermal) solar tower facility at Albuquerque, New Mexico. The mirrors focus sunlight on the top of the 61-meter-high tower, upon which experimental boilers will be tested. (Courtesy of Sandia Laboratories, Albuquerque, New Mexico.)

**C.
DISTRIBUTED
COLLECTORS**

The solar tower concept takes advantage of the high-concentration ratio for the incident radiation, achieving this with a large number of slightly curved reflectors that follow the sun through the use of a two-axis tracking mechanism. Each mirror must be continuously pointed in order to keep the sun's image focused on the boiler. The distinguishing feature of the central receiver concept is that all of the solar energy is collected at a single point.

An approach that is intermediate between the highly concentrating power tower concept and the nonconcentrating flat-plate collector is to use a series of distributed collectors, each of which has a modest concentration ratio (perhaps 10–100) to collect the solar energy by heating a fluid to some moderately high temperature, usually in the 100 to 500°C range. The basic idea involved here is certainly not new, having been utilized by a number of others in the early part of the twentieth century, such as the Shuman and Boys power plant in Egypt, which developed 50 hp for an irrigation project. However, these early engines all suffered from very low overall efficiencies of the order of 3% or less.

Modern technology offers the possibility of a number of improvements over earlier solar designs. Moderate concentration ratios can be obtained through the use of well-constructed parabolic reflectors. These can either track the sun and focus the sun's rays upon a stationary receiver or, alternatively, a stationary reflector can be combined with a movable receiver. Losses in the receiver can be minimized through the use of modern techniques such as an evacuated inner element combined with a selective surface. This would eliminate convective losses and greatly reduce radiative emission in the infrared from the hot fluid. In either case, a single-axis tracking mechanism is utilized.

Distributed collectors operate by collecting sunlight and converting it to heat at each individual collector module. Some of the common distributed collector concepts are illustrated in Figure 10-9. Each module has its own heat collection pipe (equivalent to the central receiver in the tower approach). Collection areas are typically a few square meters, so that many thousands of these modules would be required to provide the power output equivalent to a 100-MW solar tower plant. The large amount of interconnected plumbing required appears to rule out the distributed collector approach for very large installations. For small electrical power plants and for systems providing both power and heat, the distributed collector concept is becoming increasingly attractive.

A system to convert the solar energy collected into electric power is shown schematically in Figure 10-10. The concentrated sunlight is used to heat a fluid in the receiver. The combination of the concentrator, the receiver, and the associated pipes, pumps, valves, etc., is called the collector module. The heated fluid is circulated into a boiler where it heats water to form steam. In turn, the steam drives the electric generating turbine. The spent steam coming out of the turbine is cooled and condensed back to water to complete the cycle.

Most of the distributed collector designs operate on the direct component of the incoming solar radiation and must track the sun across the sky. Since the distributed collector modules have only a single axis motion, accurate aiming of the concentrator at the sun is important. Good sun-tracking subsystems should be sufficiently accurate to keep the pointing error equal to or less than 0.10°. This type of precision will keep losses due to misalignment below 2%. The automatic tracking system should include the capability to quickly detect a loss of insolation. Also, it must be able to reposition the collector, at the end of the day, to the proper orientation for collecting solar energy the next morning.

Test data for one type of distributed collector system are shown in

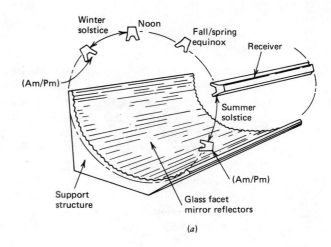

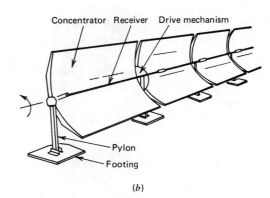

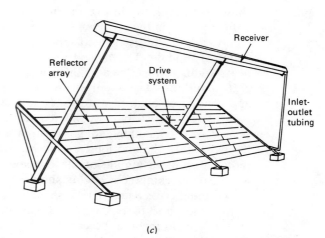

FIGURE 10-9 Some of the distributed collector concepts now under study. (*a*) Fixed stepped mirror concept. (*b*) Parabolic trough concept. (*c*) Segmented mirror concept.

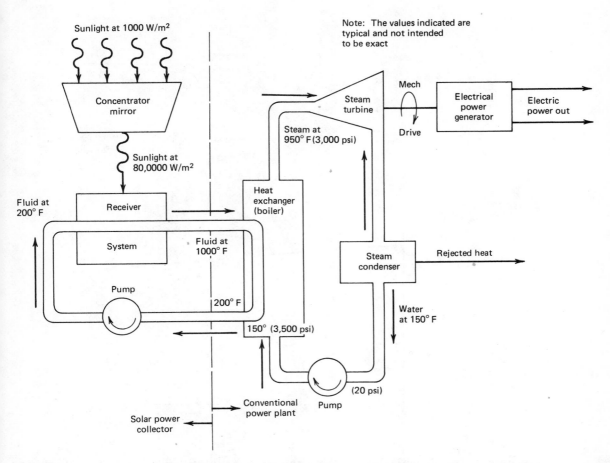

FIGURE 10-10 Schematic drawing of a possible system to generate electric power using the intermediate temperature, distributed collector approach.

Figure 10-11. The data show that good collector efficiencies can be obtained, provided that the thermal losses at higher operating efficiencies are minimized. The theoretical curve was calculated assuming a selective surface with an infrared emissivity of 0.10. These data emphasize the importance of low emissivity in reducing the thermal losses due to infrared reradiation.

A system of distributed collectors lends itself readily to a wide variety of low-power, on-site applications. This type of system can pump water for irrigation, produce steam for industrial processes, generate electricity in small- and medium-sized installations, and also supply heat for residential use. These systems may prove most economical when sev-

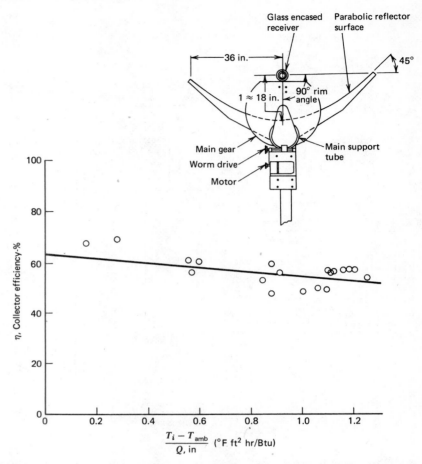

FIGURE 10-11 Summary of results obtained for a typical distributed collector using a parabolic concentrator, and a selective surface on the receiver. Insert shows details of the design.

eral purposes are served in a complementary manner. One particularly attractive application is to provide process heat for the food, textile, and chemical industries. Several systems now being studied are illustrated in Figure 10-12.

Most of the equipment needed for such systems is now available. Development of concentrating collectors that can raise fluid temperatures above the boiling point of water has progressed particularly rapidly, at a cost generally comparable to that for simpler, flat-plate collectors. The design aspect most in need of further developmental work is that for small, heat engines to convert solar heat into shaft power to drive a generator, air-conditioning compressor, or water pump. The cost of one-axis, tracking collectors is now $50 to $200 per square meter as compared to the present cost of $500 to $1000 per square meter for the more complex, two-axis tracking collector systems. The custom-made heat engines used today cost about $1000 per kilowatt, but mass production techniques should reduce this to a figure below $200.

One of the first intermediate-temperature, solar power projects to get under way was for pumping water. Northwestern Mutual Life, which has a large farm at Gila Bend near Phoenix, Arizona, undertook a 38-kW (50 hp) irrigation project in collaboration with Battelle Memorial Institute. The project was initiated in August, 1975, and the system

FIGURE 10-12   Solar Systems Test Facility at Sandia Laboratories showing the solar collector fields designed and fabricated by Suntec Systems, Inc. (right foreground) and by General Atomic Company (right background). The Suntec collector arrays have mirrored slats that track the sun and concentrate solar radiation onto a fixed receiver assembly. General Atomic's collectors have a stationary precision-cast concrete trough with silvered glass facets that focus the sun's rays on a sun-tracking receiver tube. The sun-tracking collectors on the left were designed and fabricated by Sandia. (Courtesy of Sandia Laboratories, Albuquerque, New Mexico.)

FIGURE 10-13   The solar-powered irrigation project at Willard, New Mexico. The parabolic solar collectors are shown in the center of the photograph, and the thermal storage tank and heat engine are on the left side of the collectors. The storage pond is in the foreground. The system is used to irrigate 40 hectares of different crops. (Courtesy of Sandia Laboratories, Albuquerque, New Mexico.)

began pumping water 18 months later. The system has 550 m² of collector surface shaped in the form of parabolic troughs made of aluminized Mylar. The operating temperature is about 150°C, which is considerably higher than that attainable with flat-plate collectors. At the peak of solar insolation in June, it can pump 10.6 million gallons of water per day. The overall system efficiency is between 7 to 9%. Battelle and Northwest Mutual Life have found that there are over 300,000 irrigation pumps in use in the Western United States, most requiring about the same power as the Gila Bend Facility. Another solar irrigation project is shown in Figure 10-13.

The efficiency of a solar energy system is particularly important because it is a prime factor in determining the amount of collector surface needed. In a typical solar installation, the collector cost is roughly 50% of the cost of the total system. The key advantage of intermediate-temperature systems is that they can achieve markedly better efficiencies than low-temperature, flat-plate collector systems for many applications.

D. OCEAN THERMAL ENERGY CONVERSION

In marked contrast with the above proposals are schemes that would take advantage of the small but important difference of temperature between the surface water of tropical oceans and the cooler water at a depth of 300 to 600 m or more below the surface. The ocean's thermal resource is enormous because of the distribution of water in the tropical regions. About 50% of the solar energy intercepted by the earth falls between the Tropic of Cancer and the Tropic of Capricorn. In this area, 90% of the earth's surface is covered by sea, so that 45% of all the earth's incoming solar energy is absorbed by the tropical oceans. This vast reservoir of water contains enough to supply over a thousand times the world's energy needs. The surface layer, which is a few hundred feet thick, absorbs the energy from the sun and stays at about 25°C. Cold water from the poles slides into the depths of the oceans and slowly moves toward the equator, providing a cool reservoir at a temperature of 5°C and a depth of the order of 600 m. Generally, the cold and hot layers do not mix, thereby providing a warm heat source and a cool sink for running a thermal engine.

A large fraction of the cost for most solar electric schemes is involved in the construction of the apparatus to collect the solar energy. In addition, large and expensive arrangements for thermal storage will be required. Power plants that take advantage of the ocean thermal gradients are projected to be very cost competitive with more conventional power plants because the sea acts as the medium for both the collection of solar energy and the storage of energy. The major fraction of the cost arises from the cost of construction of the thermal engine and from the need to transmit the resulting electrical energy from the plant at sea to the shore.

The idea of using ocean thermal gradients to generate electrical power is not a new one, having been first suggested by the French physicist D'Arsonval in 1881. In 1926 the French engineer Georges Claude designed a 60-kW turbine, which worked using a difference of only 20°C in the working fluid. In 1930 he attempted to build a 40-kW power plant in the tropical sea waters off Cuba. After numerous difficulties, the plant did work in a fashion, producing 22 kW from a test engine operating on a 14° temperature gradient, although whether this was a net gain of power for his plant is not clear, since energy was required to pump water up from the depths. Claude used water as the working fluid.

Because of the small, 20°C temperature differential provided by the ocean, the maximum possible efficiency would be about 6%. In fact the operating efficiency would be closer to 3% because of this 20°C temperature differential: commonly, only a 10°C differential can be used by practical heat engines. The remaining temperature difference is needed to drive heat from the warm, surface water into the heat engine and then from the heat engine to the cold, deep water.

The low-conversion efficiency of the solar sea power plants leads to a serious problem. Since the thermal efficiency of the system is less than one-tenth that of a conventional fossil-fuel plant, the heat exchanger within the boiler must transfer more than 10 times as much heat in order to obtain an equivalent power output. This implies that the boiler must be 10 times as large and 10 times as expensive. However, two other factors must be considered that tend to overcome this cost increase. The boiler tubes in a conventional power plant operate under the doubly adverse conditions of high pressure and temperature, requiring thick walls made of expensive materials. By contrast, the sea power plant boiler operates at low temperatures and pressure, so that much thinner boiler tubes can be used. In a submerged system, the sea water's hydrostatic pressure would largely compensate for the vapor pressure of the working fluid. With ammonia, the vapor pressure of the boiler would be balanced by the ocean's hydrostatic pressure at a depth of 60 m. In other words, by being able to install the boiler and condenser underwater at convenient depths where water pressure on the outside can equalize internal pressures, construction can be relatively "flimsy" with thin-walled tubes throughout.

The operation of such a sea power plant can be illustrated using the schematic shown in Figure 10-14. Warm water is pumped into the boiler to boil the working fluid, shown as ammonia in this case. The ammonia gas under "high pressure" is fed into a turbine-generator and is discharged at "low pressure" into the condenser, which also receives cold water from the deep ocean. Ammonia liquid at low pressure from the

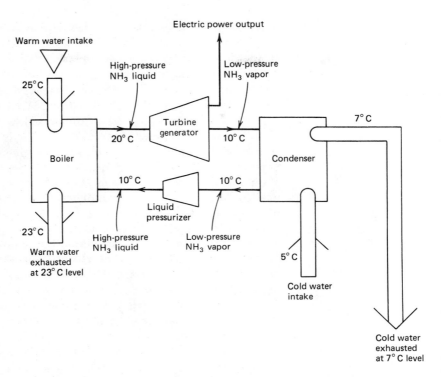

FIGURE 10-14   Schematic arrangement of a solar power plant operating on the small ocean temperature gradients existing in tropical waters.

condenser is then pressurized and pumped to the boiler where the cycle is then repeated.

Many problems remain to be solved before practical power is achieved. These include fouling of the heat exchangers by marine organisms, mooring versus dynamic positioning, which is the best working fluid, corrosion of the metal parts, etc. Those working in this area are quite enthusiastic about the possibilities of this approach because it is a field that does not appear to require any major technical developments.

E.
ENERGY STORAGE

A suitable system of energy storage is perhaps more crucial to economic solar energy development than for conventional approaches. Since the sun only shines for a portion of each day, a means to provide short-term storage of energy is an essential ingredient in any solar collection system. The common occurrence of inclement weather, particularly in northern climates, demands some type of longer-term storage. In this section we briefly review several approaches to energy storage.

The most prevalent approach to energy storage involves the storage of thermal energy as sensible heat, that is, heat is stored by raising the temperature of a given mass of material. We have already discussed the use of water and rock for heat storage at low temperatures in Chapter VIII, where it was noted that on a weight basis, water could store about five times as much heat as rock for a given ΔT. Because rock, or concrete, is considerably denser than water, there is only a factor of about two difference between the two types of storage on a volume basis. As pointed out earlier, rock storage has the advantage of allowing more stratification of the heat, thus permitting more efficient extraction of heat from the storage medium. For high-temperature thermal storage, water becomes much less attractive because of its low boiling point. Since the ability of iron oxide to store sensible heat is about 90% that of water on a volume basis, this may be a more attractive approach for high-temperature thermal storage of heat. The disadvantages of sensible heat storage systems for large-scale systems are that the storage medium takes up a lot of space, and the heat cannot be stored effectively for long periods of time.

Another method of thermal storage is to make use of the latent heat associated with a phase change in a material, such as the transition from a solid to liquid phase. For example, it takes 540 cal to change 1 gram of water from a liquid to hot steam. Unfortunately, gaseous mixtures are usually difficult to handle so that only materials undergoing a phase change from a solid to a liquid have been given practical consideration. In general, phase changes from a solid to a liquid require less energy than for a change from a liquid to a gas. The heat of fusion of sodium chloride, which undergoes a phase change near 800°C, is 123 cal/gram. The advantage of this type of thermal storage over sensible heat storage is that a great deal more heat can be stored in the same volume of material. Numerous difficulties with this approach remain to be overcome. The materials are usually expensive and corrosive. Some materials have short lifetimes due to decomposition when thermally cycled. Problems associated with supercooling have plagued the use of these materials in low-temperature applications, as the amount of heat that can be stored decreases after the heat-of-fusion salt has been thermally cycled a few times. For higher-temperature applications, eutectic salts consisting of a mixture of materials (e.g., sodium nitrate and sodium chloride) may prove useful.

Storage of mechanical energy by pumping water to a higher elevation is a technique widely used by the power industry to store off-peak power in the eastern part of the United States. The stored energy is then released at peak demand hours. The energy required to pump the water uphill is typically recovered with an efficiency of about 70%. However, the number of sites available for this approach is limited. On a large scale this would probably require moving too much water to be practical. Perhaps a better method of mechanical storage involves the use of rotating flywheels. The stored energy in this approach can be increased either by increasing the mass of the flywheel or by increasing its rotational velocity. The principle of the flywheel has been known for thousands of years, but their use has been limited by the fear of flywheel failure at high-rotational speeds, resulting in large chunks of material rocketing off and causing great damage. The real future of flywheels will depend upon the development of flywheels made up of many long, thin fibers. These flywheels would be able to withstand very high-rotational velocities, and would have the nice property of disintegrating into dust in the case of a flywheel failure.

Storage of energy by chemical batteries has always appeared as an attractive possibility. To be useful in connection with solar energy storage, the battery needs to have a large storage capacity and should have a long lifetime under operating conditions involving a large number of charge-discharge cycles. Presently the best batteries can stand less than 1000 charge-discharge cycles, whereas a lifetime characteristic of 10,000 cycles is probably needed. It is presently believed that a battery that could deliver a specific energy of 220 W-hr/kg, has a specific power of 55 W/kg, and a storage efficiency of 70% would be attractive for large-scale energy storage. One attractive possibility undergoing research now is the sodium-sulfur battery. For efficient operation it must be operated at about 300 to 350°C. The unique feature of these cells is the electrolyte, which is a solid, ceramic material called beta alumina.

We conclude by noting that the possibility of operating under a "hydrogen economy" has received a lot of attention in the past few years. The hydrogen (and oxygen) would be generated by electrolysis when an electric current is passed through water. The hydrogen could be stored or transported to appropriate sites by pipelines, and recombined with oxygen as with fuel cells; an efficiency of 50 to 60% for recombining hydrogen and oxygen to produce electricity is possible.

BIBLIOGRAPHY

1. Kenneth W. Ford, *Basic Physics,* (Waltham, Mass.: Gin and Blaisdell Publishing Co., 1968).

    An elementary discussion of thermodynamics can be found in numerous introductory texts on science and engineering. However, the presentation by Ford is extremely thorough and penetrating. Even someone with an advanced knowledge of the subject can benefit from the original and thoughtful discussion given in this textbook.

2.   A. F. Hildebrandt and L. L. Vant-Hull, "Power with Heliostats," *Science, 197,* 1139 (September 1977).

A good technical description of the solar tower concept as envisaged by the University of Houston group.

3.   C. Zener, "Solar Sea Power," *Physics Today, 26,* 48 (1973).

Heat engines operating in the tropical oceans, capitalizing on the temperature differential between upper and lower levels, could provide a source of economical, pollution-free electricity. The entire plant would be neutrally buoyant, submerged at a depth of about a hundred meters.

4.   "Energy Storage: Parts I and II," *Science, 184,* 785, 884 (1974).

An excellent up-to-date, general review of the many different storage possibilities. Contains a discussion of pumped storage, various storage battery possibilities, magnetic energy storage, mechanical storage in flywheels, chemical storage in hydrogen, and thermal storage.

PROBLEMS

1.   One of the following statements about the second law of thermodynamics is not true:
   (a) It allows one to tell how much heat is converted to useful work.
   (b) It is related fundamentally to the conversion of energy between ordered and disordered motion.
   (c) It states that one cannot run any real engine unless there is a finite temperature difference across it.
   (d) It is basically concerned with the definition of the kinetic energy of random motion of molecules.

2.   The efficiency of any real engine bears what relationship to that for an ideal engine?
   (a) Greater than.
   (b) Equal to.
   (c) Less than.
   (d) No relationship.

3.   Which of the following electric power generating plants has the greatest efficiency of conversion of stored energy into useful electricity?
   (a) A hydroelectric power plant.
   (b) Nuclear reactors.
   (c) A coal-fired power plant.
   (d) An oil fired power plant.

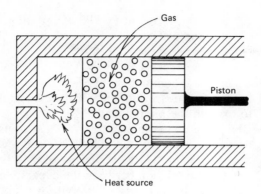

Consider the system of heat source, gas, and piston, shown above. Refer to this figure in answering the next two questions.

4. If the piston is held fixed the work done upon the piston is:
    (a) Zero.
    (b) Equal to the input heat.
    (c) Equal to the change in internal energy of the gas.
    (d) Impossible to calculate.

5. Suppose that when the piston is held fixed, the temperature of the gas increases from T_1 to T_2 when a certain amount Q of heat is added to the gas. Consider next the case where the heat Q is added to the gas, initially at temperature T_1, where the piston is free to move. The gas temperature rises to a new value T_3. Compare T_3 with T_2.
    (a) $T_3 < T_2$
    (b) $T_3 = T_2$
    (c) $T_3 > T_2$
    (d) We have insufficient information to compare the two temperatures.

6. What is wrong with using an open refrigerator to cool a house?
    (a) The cooling capacity is too low.
    (b) Nothing.
    (c) The cold reservoir is at too low a temperature.
    (d) The hot and cold reservoirs are at the same temperature.

7. In the distributed-collector solar electric scheme the magnification (concentration) factor will be of the order of:
    (a) 1
    (b) 5
    (c) 50
    (d) 1000

8. In the "power-tower," solar electric approach the magnification (concentration) factor will be of the order of:
    (a) 1
    (b) 5
    (c) 50
    (d) 1000

9. One of the following is not a method of energy storage for solar electric plants.

(a) Solar cells cooled by running water.

(b) Pumping water to an elevated reservoir.

(c) Batteries.

(d) Large flywheels.

10. Using some common, everyday examples of heat transfer, explain how difficulties would arise if the caloric theory of heat were used to explain the phenomena.

11. List three common examples of heat transfer.

12. The heat capacity of water is 1.0 cal/g-°C, while that for rock is about 0.2 cal/g-°C. If the density of rock is 2.5 times that of water, what is the relation between the capacity of water to hold heat per unit volume as compared to that for rock?

13. In Figure 10-2 suppose that the heat source is replaced with perfectly insulated material. If the piston is now moved inward so that the gas is compressed, what can you say about the temperature of the gas?

14. If the piston in Problem 13 had been allowed to move outward, increasing the volume of the gas, what could be said about the temperature of the gas?

15. In which case are the molecules of a lump of coal more ordered—before or after burning? Explain.

16. With what theoretical efficiency can the energy of a stiff breeze be converted to electrical power? Is wind an example of ordered or disordered motion?

17. Explain why the change of internal energy in a thermodynamic cycle is zero.

18. List the advantages and disadvantages of enclosing the tracking heliostats for the power-tower scheme in a plastic bubble as proposed by Boeing Aerospace Corp.

19. What are the principal advantages of the power-tower approach to solar electricity over the distributed collector arrangement? Are there any disadvantages?

20. Explain how a flywheel design, where the wheel is constructed out of fibers pointing in the radial direction, could result in greater allowable rotational velocities than with a conventional design where the wheel is constructed out of a solid piece of metal.

Photovoltaic Conversion

XI

Photovoltaic solar cells were used to convert sunlight directly into electricity as early as 1954. However, the prohibitively high cost, roughly $10 million per peak kilowatt of power, ruled out their use for all but the most exotic applications. Early technical advances in the photovoltaic conversion field resulted primarily from experience gained from the use of solar cells to power satellites and other craft sent aloft by the National Aeronautics and Space Administration (NASA). For the space effort, reliability and efficiency were the major objectives. As a result, the array cost of 1965 was still over $1 million per peak kilowatt. In this application, the solar cell has proven practical and reliable. Some solar cell systems have lasted over 10 years in space and over 15 years on the ground without serious deterioration.

The primary advantages of using solar cells to generate electricity are as follows.

1. There are no moving parts to wear out.

2. Most solar cells are made from silicon, one of the most plentiful materials on earth.

3. The overall conversion efficiency to electricity is in the 10 to 20% range.

4. The semiconductor industry is a well-developed, high-technology industry. It is probable that solar cells can readily be adapted to mass-production techniques.

The major disadvantage is the cost. In 1973, the cost per peak kilowatt of power from solar cells was still approximately $300,000. As with so many other aspects of federal policy, the 1973 Arab oil embargo caused the government to change its outlook on photovoltaic energy conversion. As a result of increased federal expenditures for photovoltaic development (federal expenditures in this area for 1977 reached nearly $60 million), the price per peak watt of power from solar cells began to drop dramatically. Solar cell array costs of $15,000 per peak kilowatt were achieved in 1977. Although not economically competitive with other forms of electric generation, this still represents a dramatic reduction in cost from the early days of the photovoltaic effort.

Energy department officials are hopeful that photovoltaic power will be competitive with other forms within the next 30 to 40 years. Specific plans are to achieve a cost of $500 per peak kilowatt by 1986. How can this large reduction in the cost of solar cells be achieved? One way would be to accomplish a major breakthrough in the research and development program. As discussed below, work is proceeding in the promising areas of thin film and amorphous silicon solar cells. Nevertheless, significant reductions in the manufacturing costs of the common, single-crystal solar cell can probably be made. For example, a reduction in cost to a factor of about 3 is possible through improvements permitting a lower grade of silicon to be used for the starting material. It has been estimated that another factor of 10 could be obtained through improvements in the methods of growing single-crystal silicon. Finally, automatic fabrication of solar cells might lead to cost savings of another factor of 20. A major difficulty of the solar cell industry of today is that it is small and production involves a lot of hand work. It should also be noted that cells made with present techniques have to operate about 12 years just to regain the energy expended in their manufacture. A major reason for the high manufacturing cost is that the raw silicon must be melted and remelted so many times during purification and crystal growth. The same studies estimating that large reductions in manufacturing costs can be achieved through technological improvements indicate that a tenfold decrease in the energy cost for manufacturing solar-grade silicon is feasible.

In this chapter we consider the conversion of solar energy to electricity with the aid of thermionic and photovoltaic devices. The use of the former is included for completeness, even though the generation of electricity by thermionic devices does involve a thermal process whose efficiency is thus limited by the second law of thermodynamics. The discussion of the direct conversion of solar radiation to electricity is broken into two parts. The first is an elementary presentation about the theory of photovoltaic cells, and the second outlines some of the promising technological developments now being pursued in the quest for an inexpensive and efficient solar cell. We conclude with a brief discussion of the proposal to put a large array of solar cells into space, and beam the collected energy back to earth.

A. THERMOELECTRIC DEVICES

It is possible to convert solar energy directly into electrical energy without using large mechanical devices involving moving parts. One way to accomplish this conversion is based upon a highly important discovery made by T.J. Seebeck in 1821. He found that if a metallic circuit composed of two different metals connected in series has different temperatures at the various junctions of one metal with the other, a voltage difference is produced; if the circuit is closed an electric current results. This type of device, called a thermocouple, has long been used as a means of measuring temperatures.

A typical arrangement for the thermocouple is shown in Figure 11-1. One junction will be heated in some manner while the other junction will be maintained at a constant temperature. Because of the Seebeck effect, a voltage difference occurs between the two junctions that is a

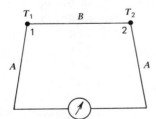

FIGURE 11-1   Illustration of the principle of the thermocouple. Junction 1 is heated to a temperature T_1, and junction 2 is cooled to temperature T_2. A resultant voltage is developed between the two junctions. Wires of two different materials are labeled by A and B.

result of the temperature difference, and an electrical current can be made to flow. If one junction is heated by solar radiation, the conversion to electricity is obtained. A larger current can be obtained by connecting a number of thermocouples in series. For effective conversion of solar energy into electricity the thermocouple should have a high voltage per degree of temperature difference, a high electrical conductivity to avoid loss by internal heating, and a low thermal conductivity to prevent transfer of heat from the hot to the cold junction.

Thermocouples made of unlike metals do not meet the above requirements very well. Typically voltages of 20 to 60 microvolts (μV) per degree are generated. Also, in metals the conduction of heat and electricity takes place through the same basic mechanism, so that it is impossible to obtain a thermocouple with both a low heat conductivity and a high electrical conductivity.

Semiconductors (see the next section) offer better possibilities. Voltages of up to 1000 μV per degree have been obtained. Also in semiconductors while electricity is conducted by free charges, conduction of heat is by an entirely different mechanism, which allows good electrical conductivity combined with poor heat conductivity.

Thermionic devices convert heat energy into work and thus are subject to the limitations set by the second law of thermodynamics. With an efficiency given by $(T_1 - T_2)/T_1$, practical limitations for the temperatures involved limit the maximum efficiency to 30 to 40%. Because of other limitations, practical solar converters of this type have efficiencies very much less than this. A thermoelectric generator constructed by Telkes in 1954 using bismuth and antimony as one element and zinc antimonide as the other had an overall efficiency of only 0.6%.

B. BASIC PRINCIPLES OF SOLAR CELLS

Another way to obtain useful solar energy is to convert it into electricity with photovoltaic devices like those carried by spacecraft. The discovery in the 1950s of the group of materials called semiconductors made practical solar cells possible with an efficiency much larger than for thermionic devices. Development of solar cells was greatly spurred by the needs of the space program for a reliable, long-lived source of power. The large majority of solar cells manufactured have been made of silicon (Si), although other substances are also suitable. Since the principles that govern the operation of all photovoltaic devices are the same, the present discussion centers upon the characteristics of silicon solar cells.

The photovoltaic method does not convert energy first to heat and then to electricity, but instead solar radiation is converted directly to electrical energy. To understand this process it is necessary to consider further the energy content of electromagnetic waves such as light. To appreciate how solar cells operate we introduce the concept of quantization. In 1905, physicist Max Planck showed that light must be thought of, not only as a wave, but also as discrete bundles of energy called photons. The energy of each photon varies according to the frequency of the wave and is given as

$$E = hf \qquad\qquad (11\text{-}1)$$

where h is a constant (called Planck's constant). Since f varies inversely as the wavelength, the energy content of a photon corresponding to waves in the blue wavelength region is greater than that for waves in the red portion of the spectrum. It is convenient to describe the energy in terms of a unit called the electron volt (eV). One eV is defined as the energy gained by an electron when it is accelerated by a voltage difference of 1 V. In these units the energy of a photon at 300 nanometers (nm) (ultraviolet) is 4.15 eV, that of a photon of 600 nm (red) is 2 eV, and that of a photon corresponding to waves of 1200 nm (infrared) is 1 eV. This relationship between energy in electron volts and wavelength is illustrated graphically in Figure 11-2.

For a practical solar cell it is necessary that the photons of the incident radiation be absorbed in some material and liberate electrons in such a way that current will flow in an external circuit. This is most usefully accomplished with semiconductors made of silicon. Silicon has a chemical valence of 4, which means that of the 14 electrons that are normally found in each silicon atom, four are available to interact with other atoms. In a silicon crystal each silicon nucleus shares its four

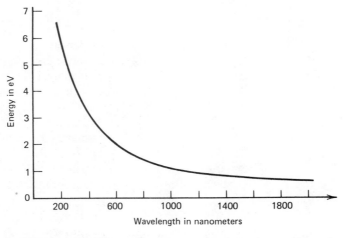

FIGURE 11-2 Relationship between the energy of solar radiation in electron volts and its wavelength.

valence electrons with four neighboring nuclei and, in turn, shares the neighboring electrons in the manner illustrated in Figure 11-3. If the structure of the silicon crystal were perfect, each electron would be held in position by the electrical forces between the electron and the two neighboring atoms who share it. Since no electrons would be free to move, a good insulator would result. This is to be contrasted with the case of metals where many electrons are free to move about so that an electric current results when a voltage is applied.

It is instructive to think of this situation in a different manner; one that can be termed a quantum description of the crystal. Figure 11-4 shows a convenient way of representing this situation. This is a band-structure diagram and is interpreted as follows. All electrons that take part in the valence-bonding process in the lattice are said to be in the valence band. If sufficient energy is provided to get them out of the valence band—that is, to break the bonding—they are then free to move and are said to be in the conduction band. The region in between is called the forbidden region or band gap. The height of this gap represents the amount of work necessary to remove an electron from the valence band and put it into the conduction band. The vertical axis represents electron energy, and the horizontal axis indicates position in the lattice (in one dimension). Electrons in the conduction band may also have kinetic energy because they are moving. Electrons with different kinetic energies are shown at different distances from the band edge. The promotion of an electron to the conduction band leaves behind a "hole"; a location that lacks an electron. The movement of a bound electron from a neighboring site into this hole constitutes a net

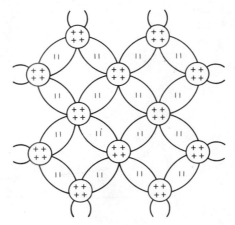

FIGURE 11-3 Covalent bonding gives rise to the regular structure of the silicon crystal. Each atom is firmly held in the crystal lattice by sharing two electrons with each of four equidistant neighbors.

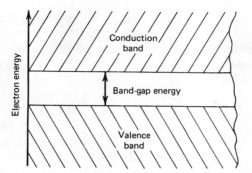

FIGURE 11-4 Schematic representation of the concept of band structure in solids. The significance of the energy gap is discussed in the text.

motion for the hole. Thus, the flow of electric current in the crystal is made up of the motions of both electrons and holes. The hole motion will be in the opposite direction to that of the electrons. This picture provides an explanation of what happens to a bound electron if it suddenly receives enough energy to escape from its bound position. One says that it has been knocked into the conduction band.

In silicon the gap energy between the valence and conduction bands is close to 1.08 eV. Only photons with an energy greater than this can release electrons from the silicon crystal. This means that only light with a wavelength shorter than 1150 nm can release electrons. Furthermore, light with shorter wavelengths will not only free an electron but also the excess energy over 1.08 eV goes into energy of motion of the electrons and is thereby eventually wasted as heat. Even photons of energy greater than 2 eV still release only one electron each. It is now possible to make a crude estimate of the maximum energy available for conversion into electricity. The solar spectrum above 1150 nm contains 22% of the total solar energy incident and this is entirely lost. Below this region, a fraction equal to 1.08 eV divided by the energy of the incident photon is recovered, since the excess over the gap energy is transformed into waste heat that only warms the solar cell. For example, the region of the solar spectrum between 500 to 700 nm contains 28% of the total energy, but only 15% can be converted into useful electrical output. If we perform this calculation over the entire solar spectrum, we find that the theoretical maximum efficiency for a perfect solar cell is roughly 45%. Any real solar cell will have further losses resulting in an even lower performance as seen below. Thus, while the limit of the second law of thermodynamics has been avoided, nature has replaced it with other restrictions making it possible to recover only a fraction of the solar energy as useful electrical output.

Why don't we get the full 45% of the incident light energy that is theoretically possible? First the mechanism for recovering the electrical energy of the hole-electron pairs formed in the junction is not perfect. Often only about one-half can be recovered, which leaves about 22%. In addition, some of the incident light is reflected at the top surface (up to 30%). One way of decreasing this loss is to coat the upper surfac with an antireflection coating. Other losses are due to absorption away from the junction region. Also, all cells have a certain internal resistance. The net result is to give practical operating efficiencies in the 10 to 20% range at present. The theoretical maximum efficiency for solar cells is shown in Figure 11-5 as a function of band-gap, and for two different operating temperatures. Lines are drawn for three common types of cells—silicon, gallium-arsenide, and cadmium-sulfide. Notice that the maximum efficiency of a silicon solar cell drops from 25% at 0°C to 14% at 100°C.

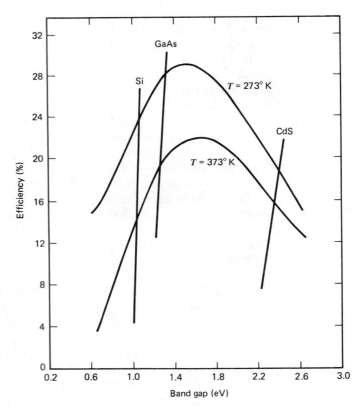

FIGURE 11-5   The maximum possible efficiency of a solar cell depends in a complicated way upon the band gap of the semiconductor material and upon the temperature.

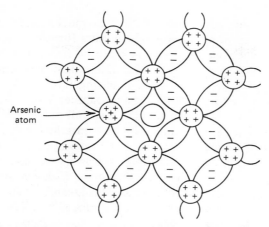

Arsenic
atom

FIGURE 11-6   The addition of a small amount of arsenic with five
valence electrons to the crystal structure of silicon
forms what is called an "n-type" semiconductor. One
electron is fairly free to move about while the stationary arsenic atom has a net positive charge.

Pure silicon is not in a form that lends itself readily to absorb photons
and give up electrons. To do this it is necessary to modify it by adding
small amounts of impurities that greatly raise its electrical conductivity.
Suppose an arsenic (As) atom is substituted for every millionth silicon
atom. Arsenic has five valence electrons so that, since only four are
needed for bonding, one electron is only weakly held by the positive
charge of the nucleus. Such a "doped" crystal is called n-type silicon,
and the bonding picture is shown in Figure 11-6. The excess positive
charges in arsenic-doped silicon are bound, while the excess electrons
move freely about. The crystal as a whole is electrically neutral, but we
now have a semiconductor with a few free electrons that can provide
an electric current under the influence of an electrical field.

It is also possible to form what is called p-type silicon by adding a
small amount of atoms with only three valence electrons, such as boron
(B). When a neutral boron atom occupies a lattice site in the silicon
crystal structure, there will be one electron missing from the usual
complete bonding. An entirely different electrical situation now prevails. A hole has been created in the lattice structure. Then, as described
above, a neighboring electron can move over and fill the existing missing
bond leaving behind a new missing bond and a net positive charge on
a silicon lattice site. This process can occur repeatedly, with the result
that electrons move in one direction and the holes in another. The
motion of the holes can be crudely regarded as being equivalent to the
motion of positive charges moving freely through the lattice. The movement of holes in the crystal lattice constitutes a flow of current.

What happens now if light strikes a bit of semiconductor material? If the photon penetrates below the surface, and if its energy is equal to or greater than the amount of energy needed to move the electron into the conduction band, it will produce a conduction electron and its associated hole. Left by themselves the electron and the hole will wander randomly about the crystal lattice. If the electron encounters a hole, the two combine with the result that the energy that had been absorbed originally to create the electron and the hole is released as heat. To make a useful solar cell it is necessary to arrange things so that the electron passes through an external circuit, doing useful work, before it finally recombines with a hole. This can be accomplished by placing two slabs of semiconductor material together, with one being n-type and the other p-type, forming what is called a p-n junction.

The trick in making a practical, photovoltaic device is to produce a p-n junction in one piece of silicon. Figure 11-7 shows a crystal in which the right-hand side has been doped with arsenic and the left-hand side with boron. Hence the right side contains a gas of free electrons and the left side a gas of free holes. Inside the junction region, diffusion causes a genuine, free-conducting electron to meet a true conducting positive hole; upon combining they neutralize each other. However, the bound positive and negative charges are left, resulting in a potential difference between the two regions. The n-type side is positive with respect to the p-type side. This electrical field will prevent the electrons released by the incident solar radiation from dropping back into the holes from which they originated.

A practical solar cell might look like the drawing in Figure 11-8. The incident solar radiation passes through the layer of p-type silicon and strikes the junction region. The released electrons eventually flow through the external load doing useful work. If absorption of light can

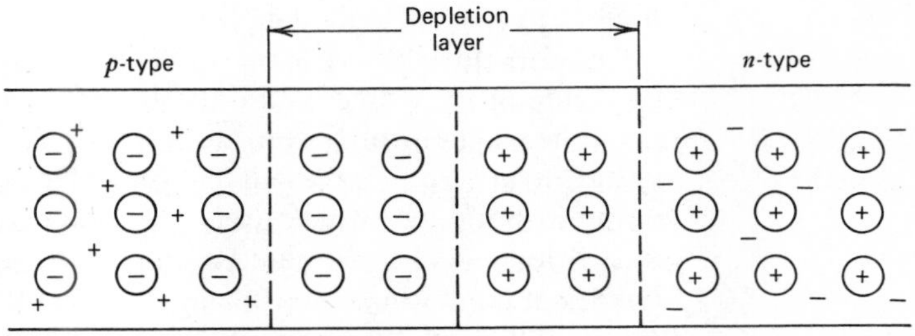

FIGURE 11-7 When p-type and n-type semiconductors are placed in contact a junction region is formed with practically no free charge carriers. An electric field is set up in the depletion region that sweeps out the electrons and holes (in opposite directions).

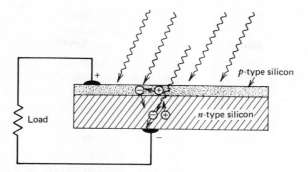

FIGURE 11-8 Cross-sectional view of a solar cell. Widely used silicon cells convert more than 10% of the incident light energy into electricity.

be made to take place within a reasonable distance from the electric field of the *p-n* junction, the electrons will be drawn to the *n*-type material and the holes will be drawn to the *p*-type material. If the terminals to the cell are connected, current will flow in the external circuit as long as the junction is illuminated. For a practical cell it is important to have a large junction area so that the flat arrangement shown in the figure is used. Also, the *p*-type layer at the top should be as thin as possible in order to minimize nonuseful absorption of solar energy. A novel way of doing this is to diffuse boron at high temperatures into *n*-type silicon forming a thin-surface, *p*-type layer. The *p-n* junction is made over the entire sample so close to the surface that light can reach the barrier through the *p*-type material.

C. THE PHOTOVOLTAIC EFFORT

The manufacturing process used today is primarily responsible for the high price of solar cells. It is worthwhile to outline briefly the process presently used to manufacture silicon solar cells. The jumbled, polycrystalline structure of purified silicon is first converted into an orderly structure of single-crystal silicon by what is known as the Czochralski growing process. This provides single-crystal silicon in the form of cylindrical ingots two or more inches in diameter. To prepare an ingot, a batch of silicon is melted and minute amounts of boron or phosphorus added. Then, a silicon seed crystal is lowered into the 1420°C liquid, and an ingot is "grown" by twirling and pulling the seed upward at a few inches per hour. If the temperature and the rates of rotation and pulling are controlled carefully, a perfect crystal will be formed. Once an ingot is grown, up to three-quarters of it is destroyed in the remaining production process. A circular diamond saw cuts thin slices about 25 to 30 thousandths of a centimeter thick. Since the saw blade is also this thick, much of the ingot becomes sawdust, thereby raising the cost. Grinding and polishing these silicon wafers pushes labor costs higher. The wafers are then baked in a chemical atmosphere that diffuses another semiconducting layer into one surface. Finally, electrical terminals are attached, and the wafers are wired into solar panels. To summarize, the principal reason for the high cost of solar cells is that the growth, slicing, and fabrication processes are all highly labor intensive, requiring skilled workers. Automating some steps should bring costs down to less than $5000 per kilowatt in a few years, which is still too expensive for large arrays.

One promising way to lower the crystal-growing cost involves the growth of continuous ribbons of silicon by a process known as edge-defined, film-fed growth (EFG). Originally, EFG was developed to speed the growth of single-crystal sapphire used for electrical microcircuits. When a die is lowered into molten silicon or sapphire material, the liquid rises through its center by capillary action. The material flows only to the top edge of the die, whose shape determines whether a tube, square, ribbon, etc. is formed. A seed crystal pulls the solidifying material upward; one thin dimension must exist to enable heat to dissipate quickly. Ribbons several feet in length have been pulled from a crucible of molten silicon in one hour. The silicon crystals grown by this EFG process are structurally far less perfect than those grown by the Czochralski method. Solar cells with efficiencies between 10 to 12% have been constructed from the EFG material, but the process introduces unwanted impurities into the silicon and is not yet as rapid as the traditional method.

Although the direct use of silicon solar cells has received the most attention so far, a number of other possibilities are being studied. One attractive idea is to use concentrating, photovoltaic systems. Systems presently under consideration cover a range of concentrations from 10 to over 1000. A variety of innovative concentrating collectors have been designed. With one exception, the concentrating systems will be sun tracking. The exception is a trough-shaped reflector known as the Winston collector. This device has the nice property of focusing all incident radiation within a certain range of angles upon a small amount of surface area. A concentration factor of about 10 can be utilized without the need to track the sun across the sky. At high concentrations most collectors will require cooling, because the performance of solar cells goes down as the temperature of the device increases. A tremendous advantage of concentrating systems is that economic studies show that the cost of generating electric power is almost independent of the cost of solar cells, provided that the latter is below a critical value of about

$1000 per square meter. One example of a concentrating system is shown in Figure 11-9.

It is generally felt that concentrating photovoltaic systems will need to utilize very high-efficiency solar cells. Because of this, interest is high in gallium-arsenide photovoltaic cells, despite the fact that they are presently 10 times as expensive as their silicon counterparts. Varian Corporation has demonstrated an experimental system with an efficiency of 19% when operated with a concentration factor of 1935, while IBM has tested a gallium-arsenide cell having an efficiency of 22%. A key advantage of gallium-arsenide cells for this application is that they can tolerate higher temperatures than silicon, up to 200°C with only a moderate loss in efficiency. Still to be answered is the question of

FIGURE 11-9   Using 135 plastic fresnel lenses mounted directly in front of the corresponding number of silicon solar cells, the photovoltaic concentrator array pictured above produces 1 kW of electricity. Each fresnel lens, which concentrates the sun's rays like a magnifying glass, focuses the equivalent of 50 suns on a corresponding solar cell to convert sunlight directly to electricity. This high-intensity illumination increases each solar cell's electrical output to reduce the cost of the resulting electricity. (Courtesy of Sandia Laboratories, Albuquerque, New Mexico.)

whether or not a sufficient resource of gallium is available to permit the large-scale use of such cells.

Another approach involves the use of solar cells made from thin films of polycrystalline or amorphous semiconductor materials. Optimism about this possibility is based on the large cost savings possible due to reduced material demands and manufacturing effort. Thin cells on the order of a few millionths of a meter can be formed quickly using chemical deposition or spray techniques. While thin-film polycrystalline solar cells made from silicon would be cheap compared to those manufactured in the conventional way, their low efficiencies of 1 to 2% would require prohibitive amounts of surface area. Cells made from thin films of cadmium sulfide (combined with copper sulfide and other compounds) have also been intensively studied. These cells can be produced by vacuum depositing materials on plastic, a process that lends itself readily to mass-production methods. Efficiencies in the range of 5 to 8% have been achieved with these devices. However, their performance appears to degrade rapidly when exposed to air and under elevated temperatures. Perhaps an even more attractive possibility is to use cells fabricated from an alloy of amorphous silicon and hydrogen. The hydrogen, which can be added in varying amounts up to about a one-to-one atomic ratio, acts to increase the absorption of light in the film along with improving its photovoltaic properties. Efficiencies as high as 10% are expected within a few years.

The dominant line of thought at present in the federal research and development program is that photovoltaic devices can have a major impact as an energy source only if large, utility-type applications can be found. Photovoltaic systems are inherently modular. The basic unit for these solar systems is an array of cells producing up to a few tens of kilowatts. Large-scale installations would be constructed by putting together many such modules. But, there is no technical reason why they cannot compete with other sources of electricity on all scales. One of the goals of the present federal solar-electric program is to have a 10-MW photovoltaic demonstration plant in operation by the early 1980s, with solar cells making a 1 to 2% contribution to the country's energy budget by the turn of the century. Only time will tell if this represents an optimistic or pessimistic point of view.

D. ORBITING SATELLITE SYSTEM

If the cost of solar cells continues to decrease as predicted, both small and large arrays of photovoltaic devices will soon become a practical eventuality. One very intriguing usage of large solar-cell arrays is to place them in orbit above the earth. In this case the problem of irregular collection due to clouds would be eliminated. Storage of energy would no longer be a problem, since a single satellite would view the sun for

23 out of every 24 hours. In addition, the intense, almost constant solar flux reduces the collector area requirement by almost a factor of 8. It is easy to calculate the area required in space needed to provide the U.S. energy consumption of 1970 (6×10^{19} J) assuming an overall conversion efficiency to electrical power of 15%. The answer is that an area of about 9500 km^2 would be needed, which is one-eighth as much as would be needed for a surface location. On a smaller scale, the area needed for a 1000-MW plant would be under 5 km^2.

Spurred on perhaps by the success of the U.S. satellite program, Leon Gaudrer of Texaco made the first serious suggestion for the use of an orbiting power satellite in 1965. In 1968, Peter Glaser of the Arthur D. Little Company suggested a detailed proposal for the large-scale production of power in space. A large satellite, such as that shown in Figure 11-10, would be placed into a synchronous orbit of the earth, remaining fixed above a given ground location. The satellite would consist of a solar collector, an energy conversion device to provide electric power, a transmission cable to supply the electricity to micro-wave generators, and an antenna arrangement to beam the microwave radiation to a receiving station on earth.

The geometrical arrangement for the satellite system is shown in Figure 11-11. At an altitude of 35,500 km in an orbit parallel to the earth's equatorial plane, a satellite moving from east to west would be stationary with respect to any given point on earth. Since the satellite will lie in the earth's shadow once a day, two satellites in the same orbit, but spaced 21° apart, would be required in order that at least one would be illuminated at all times. A network of satellites that could continuously serve a large number of stations on the ground has been suggested.

While the development of economic photovoltaic devices is the single most important technical development needed, there are many other highly interesting technological problems that would have to be overcome. For example, the solar collectors should be oriented toward the sun to within about 1°. Glaser maintains that, except for system size, this is within the realm of present tracking technology. A more difficult requirement is that of locking onto an earth-based antenna. In order to keep the microwave beam from straying more than about 150 m, a pointing accuracy of 0.5 sec of arc is required. A guidance and control system will have to neutralize the effects of various forces acting on the satellite in order to maintain a stable orbit. The largest of these forces is due to radiation pressure from sunlight; small control rockets may be used to help overcome this. Cooling of the microwave generators will have to be handled by some type of closed circulating system, with the waste heat being radiated into space by means of heat pipes on space radiators that are distributed over the structure of the microwave antenna.

FIGURE 11-10 Solar electricity could be generated in space by mounting huge arrays of solar cells on an orbiting synchronous satellite. The collected energy would then be beamed back to earth in the form of microwave radiation. (Courtesy of Arthur D. Little, Inc., Cambridge, Mass.)

Present microwave technology should be sufficient to convert the electrical power to microwave radiation, say of 10-cm wavelength, with an operating efficiency of 75 to 90%. Choice of 10-cm wavelength radiation minimizes atmospheric absorption of the transmitted energy. A power density of about 1 W/cm² for the transmitted radiation has

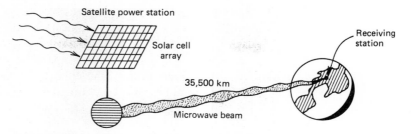

FIGURE 11-11   Illustration of the solar satellite concept.

been suggested. This is roughly an order of magnitude higher than the normal solar radiation intensity. It is sufficiently high to damage objects or living tissues that enter the beam. Regulations and controls to prevent the entry of objects or living beings into the beam would have to be imposed.

BIBLIOGRAPHY

1. D. M. Chapin, "The Direct Conversion of Solar Energy to Electrical Energy," Chapter 8, of *Introduction to the Utilization of Solar Energy;* A. M. Zarem and D. D. Erway, editors (New York: McGraw-Hill, 1963).

    A review of semiconductor principles—design of solar cells, their fabrication, theoretical efficiency, and economics.
2. B. Chalmers, "The Photovoltaic Generation of Electricity," *Scientific American* (October 1976).

    An elementary and very readable discussion of the principles of solar cells. Contains a useful summary of present manufacturing methods.
3. P. E. Glaser, "Power without Pollution," *Journal of Microwave Power, 5,* 211 (1970).

    A description of the orbiting satellite concept. The solar energy would be converted to microwave radiation and beamed down to earth. System considerations and the developmental tasks for a large-satellite, power station are reviewed and the potential technological needs are identified.

PROBLEMS

1. A difference of temperature between the two junctions of a thermocouple will produce:

    (a) A flow of holes.

    (b) Melting of the wires.

    (c) A voltage difference between the two junctions.

    (d) No measurable effect.

2. *n*-type silicon is formed by adding minute amounts of elements with the following number of valence electrons:
    (a) 3   (c) 5
    (b) 4   (d) 6

3. At room temperatures the theoretical maximum efficiency of a solar cell cannot exceed about:
    (a) 5%   (c) 45%
    (b) 15%   (d) 75%

4. Practical thermionic devices appear to have a very low efficiency for the conversion of solar energy to electricity. Can you suggest some reasons for this?

5. Explain why the maximum efficiency of real solar cells is only about 22% at the most.

6. Can you suggest something besides arsenic as the impurity for making *n*-type-silicon, semiconductor material.

7. Do the same thing as in Problem 6 for *p*-type-silicon, semiconductor material. This time you are looking for a substitute for boron.

8. Why does one often construct solar cells in the form of a flat wafer?

9. Can you think of some reason why the efficiency curves of Figure 11-5 fall off at high-band-gap energies?

10. Do practical solar cells need to be cooled? Explain.

11. In the discussion of the plan to put a large array of solar cells out into space, it was noted that the required surface area in space for the solar cells would be almost a factor of 8 less than for a similar power plant on earth. Explain how this factor of 8 is obtained.

12. Solar cells are one of the many applications of solid-state devices. In this connection it is worth noting that transistorized electronics are much more reliable than their vacuum-tube counterparts. List the reasons why this statement is true.

13. What is the key advantage of polycrystalline silicon solar cells over those grown from single crystals?

14. An array of thermocouples is often used as a quantitative detector of solar radiation. Research the available literature on this topic and write a short paper on the subject. Contrast the advantages and disadvantages of this approach with one other type of radiation detection.

Flat-Plate Collector Heat Loss **I** Analysis

In Chapter VII it was stated that the heat loss through the top of a flat-plate collector is

(I-1) $$Q = U_{top} \, \Delta T$$

Where U_{top} is an overall upward heat transfer coefficient, and ΔT is the temperature difference between the absorber plate and the outside. The absorber plate temperature can be approximated by the average of the inlet and outlet temperatures of the heat transfer medium. (Correction

257

must also be made for the nonuniform nature of the absorber plate temperature distribution.) The purpose of this presentation is to illustrate how U_{top} can be obtained from the individual heat transfer coefficients corresponding to the upward heat losses between adjacent surfaces, and to show the dramatic decrease in heat loss as the number of cover plates is increased.

The procedure will be outlined for a liquid-type collector as shown in Figure I-1. The total temperature difference between the absorber plate and the outside world can be written as

$$(I\text{-}2) \qquad \Delta T = \Delta T_1 + \Delta T_2 + \Delta T_3$$

At equilibrium, the heat loss Q between adjoining surfaces is given as

$$(I\text{-}3) \qquad Q = U_1 \Delta T_1 = U_2 \Delta T_2 = U_3 \Delta T_3$$

In this relation, each of the U's is equal to the sum of its radiative and convective components. If we substitute for the ΔT's, we have

$$(I\text{-}4) \qquad \Delta T = \frac{Q}{U_1} + \frac{Q}{U_2} + \frac{Q}{U_3} = Q\left(\frac{1}{U_1} + \frac{1}{U_2} + \frac{1}{U_3}\right) = \frac{Q}{U_{top}}$$

where

$$(I\text{-}5) \qquad \frac{1}{U_{top}} = \frac{1}{U_1} + \frac{1}{U_2} + \frac{1}{U_3}$$

The calculation of U_{top} is thus easily done once the various individual terms are known.

To conclude this discussion it is instructive to illustrate how additional cover plates reduce the overall heat loss. Suppose that without any

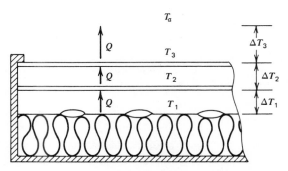

FIGURE I-1   The temperature distribution in a flat-plate collector with two cover plates.

cover plates, the heat transfer coefficient for the top is 30 W/m²-°C, a typical value for an unprotected surface in a mild wind (which causes severe convective losses). The heat loss Q would then be 30 ΔT. If a cover plate is added, the heat transfer coefficient between the absorber plate and the glazing is now about 9 W/m²-°C, while the heat transfer coefficient between the glass cover plate and the outside world is still 30 W/m²-°C. The overall heat transfer coefficient is

$$\frac{1}{U^1_{top}} = \frac{1}{30} + \frac{1}{9} = \frac{1}{6.92}$$

The heat loss from the collector is now 6.92ΔT, a reduction of over a factor of 4.

The addition of a third cover plate gives

$$\frac{1}{U^2_{top}} = \frac{1}{9} + \frac{1}{9} + \frac{1}{30} = \frac{1}{3.91}$$

The upward heat loss is now reduced to 3.91ΔT, which is a further substantial reduction over the heat loss obtained with only one cover plate. Remember, of course, that this reduction in heat loss is accompanied by a loss in the overall transmission through the additional cover plates. At higher temperatures the reduction in heat loss normally justifies the use of two cover plates.

Solar Ponds

In 1948 Rudolph Block suggested that an effective solar collector could be created by avoiding convection in a stratified salt solution. This phenomenon had been observed as early as 1900 in several Hungarian lakes. It was noted that solar radiation absorbed in the body of these lakes resulted in an increase in temperature with depth. Because of a natural salt-concentration gradient in the lakes, lower regions remained denser even when warmer. Convection is prevented because the increase in density offsets the effect of thermal expansion due to local heat absorption. When no convection can occur, water can lose heat only by conduction, and this heat-loss mechanism is slow. One meter of nonconvective water is about as good an insulator as 6 cm of styrofoam. Except right at the surface, radiation losses are zero, since water is opaque to low-frequency, infrared radiation. In natural solar ponds the salt-concentration gradient is maintained by means of salt deposits at the bottom of the lakes that provide a close to saturated solution there, and by fresh water streams that flow across the top of the lakes. Toward the end of summer, temperatures above 70°C (158°F) have been measured at a depth of 1.3 meters.

Under suitable conditions, it appears likely that even higher-temperature rises might be attained. Thus, the possibility of utilizing the solar energy collected in the form of heat at the bottom of such a pond becomes more practical. In addition to its relative simplicity, the solar pond has the attractive feature of having a long-term storage capacity built into the solar collector, in contrast with the few days of heat storage that is usually available with water tanks and rock beds in more conventional systems. This suggests its application to space heating, particularly to housing complexes where a large solar pond might provide sufficient heat for several buildings.

Experimental studies on solar ponds were carried out in the early 1960s in Israel by Tabor and collaborators. A 25-meter solar pond with a blackened bottom was constructed on the shores of the Dead Sea. Using an artificially created salt-concentration gradient, they were able to obtain a temperature of over 90°C at a depth of 0.8 m. The goal of this study was to achieve economical electrical power. In his original paper, Tabor felt that power generated using the solar pond concept might prove competitive with commercial fuels. Thus far, however, this concept has not been actively pursued.

The principle of a solar pond is illustrated in Figure II-1, where a regular pond of water that is homogeneous throughout is compared with a solar pond in which a salt-concentration gradient is maintained with the lighter, fresh water on top and the heavier, salt water at the bottom. Both ponds are sufficiently transparent that a large fraction of the solar energy penetrates to the bottom and is absorbed there. As the water at the bottom of the homogeneous pond warms up it expands and rises to the surface where it cools off again. In the solar pond the thermal expansion is insufficient to disturb the stability provided by the salt-concentration gradient and convection is suppressed. Since the heat loss upward by conduction is very slow, the temperature at the bottom rises and a permanent temperature gradient is established. Useful heat can then be extracted from the bottom of the pond, although care must be taken not to upset the stability of the solar pond.

Because the incident solar radiation covers a wide spectrum of wavelengths, only a portion of it will penetrate to the bottom of the pond. About 24% of the insolation lies in the relatively short-wavelength region of 200 to 600 nm. At a depth of 1 meter, only one-eighth of this has been absorbed by the intervening solution, so that most of this 24% will be absorbed by the bottom layers or by the blackened bottom. About 35% of the incident solar radiation lies in the wavelength interval of 600 to 900 nm; of this one-third penetrates a depth of 1 meter. The remaining 40% of the incident solar radiation is absorbed in the first 10 cm of depth because of the opaqueness of long-wavelength radiation in water. Even without detailed calculations it is evident that the optimum depth of a solar pond lies between 1 and 3 meters.

The warming of a solar pond is shown, in Figure II-2, to be a function of the time after filling. In steady-state conditions the pond follows the pattern of the incident solar energy with a lag in phase of about two to three months. This conclusion was reached in both instances as a result of both theoretical analysis and some limited experimental observations.

The problem of keeping the pond clear is a serious one, although probably simpler than trying to keep equally vast areas of glass or

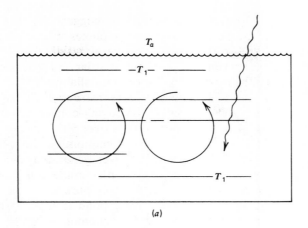

(a)

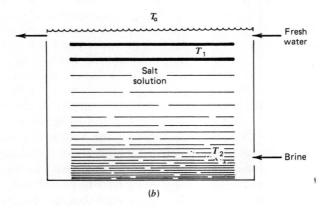

(b)

FIGURE II-1 A nonconvecting solar pond is contrasted with an ordinary water pool. The temperature of the water in the homogeneous pond in (a) is essentially constant throughout, resulting from the circulation of water. In the solar pond shown in (b), the stratified salt solution stops all convection and permits a temperature gradient to be attained with $T_2 > T_1$.

mirror clean. If the dirt consists of floating particles, these particles can probably be removed by the surface-washing process. Heavier particles that sink to the bottom will cause no problem unless they reduce the effectiveness of the black absorber on the bottom. Particles that remain suspended are more serious, since the effective transmissivity of the pond will be lowered. Biological growth can be suppressed by chemical means.

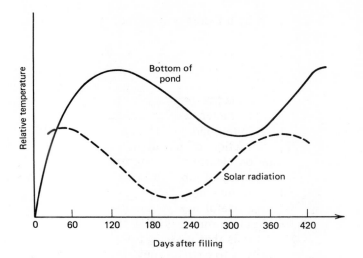

FIGURE II-2   The expected temperature near the bottom of a solar pond is shown as a function of the time after filling. Also illustrated is the phase relationship between the pond temperature and the insolation. Roughly speaking, there is a lag of about 60 to 90 days between the two quantities.

The rate of molecular salt diffusion is remarkably small. It is claimed that no action needed to be taken to correct for diffusion in the small experimental ponds. To maintain the gradient it is only necessary to inject a concentrated brine solution into the bottom of a pond from time to time. The top of the pond is washed with a weak solution with some overflow. An equilibrium concentration will be created at the top so that the salt carried away in the overflow equals the salt diffusing upward through the pond.

Whether or not the ground below the pond can provide additional storage of heat depends upon the level and flow rate of ground water and on the size, since ground storage is feasible only if the pond is large compared to the distance the water moves per year. If the ground leakage is fast, the bottom and sides of the pond should be insulated. Otherwise ground storage is an attractive feature, since it offers heat storage equivalent to nearly 1 meter of water at no extra cost.

For a small pond, the heat losses from the boundaries are excessive. As the pond is made larger the effect of the sides decreases, and the heat loss from the bottom vanishes after the pond has heated (provided that ground-water flow rates are not excessive). Weinberger has shown that a small pond gives a lower ultimate temperature rise than an infinite pond by a factor of approximately

$$(\text{II-1}) \qquad \frac{\Delta T}{\Delta T_{max}} = 1 - \frac{2h}{D}$$

where h is the depth and D is the diameter. This means that, for a pond that is 1-meter deep, the temperature rise will be within 90% of that for an infinite pond if D is larger than 20 meters.

Extraction of useful heat from a solar pond is a problem whose solution clearly awaits future experimental work. The simple solution of an array of pipes through which a heat-exchange fluid such as water is piped may not be practical for the type of pond outlined in Figure II-1. For useful heat exchange to take place, it is necessary that some convection outside the pipes occur. Nevertheless, the danger of convection is that it might destroy the stability of the pond. Another possibility is to extract liquid from the bottom of the pond, pass it over an external heat exchanger, and then return the liquid to the pond. Because of the density gradient it is known that a layer can be removed without causing mixing. The need to return the extracted fluid, which will now be well mixed, requires a homogeneous convecting layer at the bottom. A possible variation on this possibility is to construct the solar pond with a glass or plastic partition between the upper insulating layer, with its necessary salt gradient, and the bottom homogeneous layer, which is convecting.

BIBLIOGRAPHY

1. H. Tabor, "Solar Ponds—Large-Area Solar Collectors for Power Production," *Solar Energy, 7,* 189 (1963).

   Contains a detailed discussion of the problems involved and the economic aspects. Describes a number of experimental results.
2. H. Weinberger, "The Physics of the Solar Pond," *Solar Energy, 8,* 45 (1964).

   Presents a mathematical treatment of a solar pond. The efficiency of the pond as a collector of solar energy is estimated as greater than 20%.
3. F. Zangrando and H. C. Bryant, "A Salt Gradient Solar Pond," *Solar Age, 3,* 21 (April 1978).

   Describes a circular solar pond constructed at the University of New Mexico. Important references on the subject of solar ponds are listed at the end of this informative article.

Index

DATE DUE			
AP 21 '85	APR 24 '85		
AP 26 '85	APR 25 '85		
NO 9 '87	NOV 17 '87		
AP 25 '88	APR 11 '88		
MAR. 1 5 1993	MAR 8 '93		

DEMCO 38-297